WHAT LIES
BENEATH

WHAT LIES BENEATH

MY LIFE AS A FORENSIC SEARCH AND RESCUE EXPERT

PETER FAULDING

MACMILLAN

First published 2023 by Macmillan
an imprint of Pan Macmillan
The Smithson, 6 Briset Street, London EC1M 5NR
EU representative: Macmillan Publishers Ireland Ltd, 1st Floor,
The Liffey Trust Centre, 117–126 Sheriff Street Upper,
Dublin 1, D01 YC43
Associated companies throughout the world
www.panmacmillan.com

ISBN 978-1-0350-0590-1

3 5 7 9 8 6 4

A CIP catalogue record for this book is available from the British Library.

Typeset in Janson Text LT by Palimpsest Book Production Ltd, Falkirk, Stirlingshire
Printed and bound by CPI Group (UK) Ltd, Croydon, CR0 4YY

Visit **www.panmacmillan.com** to read more about all our books
and to buy them. You will also find features, author interviews and
news of any author events, and you can sign up for e-newsletters
so that you're always first to hear about our new releases.

Dedicated to Mum and Dad, to whom I owe everything and who made me who I am today. From an early age, your guidance instilled the principles of hard work, good manners, respect, high standards and strong family values in me. I am forever grateful.

In 2021 my grandson Theodore was born.
I hope this book inspires him to follow his dreams.

Contents

Glossary

Carbide lamp – a miner's lamp that produces acetylene gas by a chemical reaction between calcium carbide and water; the acetylene gas is ignited and burns with a bright white flame.

GPR – ground-penetrating radar.

HEMS – the Helicopter Emergency Medical Services.

HSE – the Health and Safety Executive.

PC – a police constable.

PJI – a parachute jump instructor in the Armed Forces.

PolSA – a police search adviser; plans and coordinates crime and missing person searches.

ROV – remote-operated vehicle, for search operations.

SGI – Specialist Group International, the search and rescue company run by Peter Faulding.

SIO – a senior investigating officer; the police officer in charge of a particular investigation.

SSS – side scan sonar; this equipment sends out sonar waves which, when they bounce back and are processed, can build a picture of a riverbed or lakebed, showing objects lying on the bottom.

Static line – a webbing strap attached to an aircraft and to a parachute; as a paratrooper jumps from the aircraft, the parachute is pulled from its backpack and is opened automatically.

Wire ladder – a lightweight roll-up ladder that cavers use to climb up and down shafts; normally ten metres long and can be joined together with other ladders for greater lengths.

Prologue

The ground I was standing on had kept its secrets for four years. And judging by the mess around me, it wasn't going to give them up easily now. Somewhere in the soil there was a body. Her name was Kate Prout. She'd been placed there by her husband who had admitted to murdering her a week previously, after which he led the police to the spot where he claimed to have disposed of her body, the place where I was now standing. It was a wooded crime scene of around 250 square metres, over which had already trundled police officers, a cadaver dog and handler, a mechanical digger, a man with a radar and a forensic archaeologist. Lots of people, but no body.

Kate was last seen alive on 5 November 2007. Bonfire night. The final contact she made was a call to her bank. Then she disappeared. Five days after she was reported missing by her husband, Adrian. They were in the process of a messy divorce. You can probably guess the rest.

Four years later, after being convicted of her murder by a jury, Adrian, who professed his innocence from jail for several months after the trial, finally caved in and confessed. Which is why I was in woodland with a team of police officers, crime scene investigators and forensic archaeologists, surveying a landscape.

My name is Peter Faulding and I find things. By the time I'm called to places like this patch of land in Gloucestershire, things have generally gone south for someone. I'm rarely asked to look for the living, I'm looking for the dead, or for the clues that will tell the authorities how they met their end. Sometimes it's simple: the body is in a garden, or under floorboards. But sometimes, when the murderer chooses open space – a field, a wood, a lake – it's much harder. X rarely marks the spot. In those cases, the police call me.

My ability to find things comes from experience and a curious mind. After so many years and so many searches, my ability to locate people and things is almost instinctive. I know where to look and what questions to ask of the landscape. Is that depression a drain, or a grave? Is that mound a tree root, or a buried torso? There are signs everywhere. You just need to be able to read them. My world is below the surface, and on that cold grey November afternoon in 2011 I scanned the scene calmly and began to digest the information before me.

I was there with three of my colleagues from the search and rescue company I run, Specialist Group International (SGI). I had been called up the previous day by the detective inspector in the case, Giulia Marogna. She was not in a good mood. The search should have been easy. Kate's body should have been located, recovered and in the morgue by now, but for some reason, even though Adrian had taken officers to where he said he had buried her, she could not be found. Instead, everywhere you looked there were holes. There was a JCB standing still to one side, ready to do any heavy work required, and several people with spades.

'We're struggling to find her – I need you to do your magic on this one,' Giulia said.

A search starts with questions. How long ago was the body buried? What type of soil or ground cover are you dealing with? Could the body have been moved after it was dumped? Could there have been a water course running over the area at any time? Are there wild animals around? All these potential variables will affect how and where the search takes you.

I crouched down and scraped the ground with the trowel I always carry with me on any searches, and knew immediately what the problem was. Radar equipment had been used that day but the soil was fine-grained and sandy, not ideal conditions for ground-penetrating radar. I had years of experience working with this type of equipment so I knew the limitations in these soil conditions. Radar shows disturbances and anomalies under the surface, but grains of sand evenly redistribute themselves after they've been disturbed by things such as rainfall. Think of how sandcastles on the beach disappear after the tide has come in, leaving no trace. After four years underground, all the soil on top of Kate would have evened itself out.

We needed to wind back and start from the only concrete thing we knew, which was that Adrian had told the police a specific area in which he said he had buried the body. Before the search operation started there were five pheasant pens in the area, simple structures of four posts with corrugated iron across the top where pheasants could shelter. Adrian admitted he buried his wife in front of one of these. They had all been dismantled by the time I arrived.

'Can we put everything back as it was?' I asked Giulia. She gladly obliged and ordered everyone off the scene so that me and my colleagues could return the area to how it was when they had arrived and start again, focusing on the location Adrian had identified. Once that had been done I briefed my team on what they needed to do the following morning as I had to attend another police operation.

Having re-established the scene, I tasked one of the team, Chris, to scan the area with our powerful metal detector, as often bodies are buried in shallow graves with jewellery still on them. Within a few minutes the detector issued a high-pitched squawk. The area was swept to double check, and again the machine beeped.

The spot the metal detector had indicated was marked and the forensic archaeologist moved in.

Time can appear to stand still at these points in a search when it's clear the target is in sight, and as the soil was methodically and slowly scraped away it was obvious what we'd discovered. Exposed human fingers jutted from the shallow hole in the earth. They were stark, almost obscene against the surrounding woodland floor, set into a claw, decomposed, with bone and tendons showing through discoloured flesh.

A hush fell over the scene. The only noise was the soft scraping of the archaeologist's trowel as they continued to reveal more of the body. Something sparkled in the light. As more soil was brushed away, Kate's wrist was revealed. There was still a watch on it.

Chapter 1

It is the late 1960s and a Halloween night in a field in Surrey by the side of a quiet road which will one day be the M25. A waxy moon sheds an eerie glow over the grass. The sound of cows gently mooing in the distance punctuates the silence.

At the side of the road in Gatton Bottom, near Merstham, a group of adults stand around a deep hole, looking down. Carbide lamps are placed around the rim, illuminating the first few feet of the square shaft dug through the rock. A wire ladder and a rope disappear into it. Beneath us, forty-five feet underground, shadows dance off the walls and a child's voice can be heard. It belongs to a nine-year-old, whose slight frame is now edging, inch by inch, down into what for all intents and purposes is a compacted earth tomb. He is secured on the rope, which one of the adults feeds down as he clambers down the ladder into the gloom. After several painstaking minutes the lifeline goes slack. Everything is still. An owl hoots in the distance.

'Peter?' one of the adults shouts.

In the ground under their feet a child had dropped through the ceiling of a forgotten ancient mine and was standing in a chamber that hadn't echoed with the sound of human feet for hundreds of years. His heart was beating with excitement, adrenaline surging, as he shone his headlamp over tunnel walls carved out by miners in the late Middle Ages.

The boy was me, Peter Faulding, and the shaft I was lowered into that night was an enlarged bore hole that had been sunk by contractors building what would eventually become London's orbital motorway and one of the most congested roads in Europe. The adults on the surface were my parents – my mum Nora, my dad John and their friends Dennis Musto, Mick Clark and Graham. While this scene would have appeared bizarre and probably macabre to any passers-by, particularly given the date, for the Fauldings this was a normal family outing. We were cavers. Dad was a highly experienced subterranean explorer and for years I had accompanied him on his weekend expeditions into the abandoned mines that stretched for miles in and around Surrey.

We were in this field because a few months before hundreds of bore holes had been sunk by the construction company contracted to build the motorway in an area that was honey-combed with disused firestone mines. Dad's friend Dennis, a local historian and caver, was acting as a consultant. He and the crew had spent many months – every Sunday morning and some weekday evenings – widening the bore hole by chiselling out the square shaft that I had descended into to see what was at the bottom of it. They used an old ex-military canvas bucket attached to a rope to haul the rubble to the surface. The 'spoil' was then emptied in a square-shaped pile around the top of the shaft.

Eager to discover what lay beneath, Dad and the team had reached the last two feet of the shaft before the hole opened into the void and eagerly chiselled away to make it wide enough for a small person to squeeze through.

While most children would be terrified at the prospect of being lowered into a deep, dark hole in the ground on Halloween, I was thrilled. When I first manoeuvred my body into the opening, it was so tight I could barely fit my helmet through, so I had to remove the light attached to it, which I carried in my hand.

As the shaft was dug, pieces of earth and rubble had dropped down the bore hole to form a heap a few feet high under the hole in the ceiling through which I had been lowered. I landed on the pile and clambered down the mound, shining my lamp around the tunnel.

'What can you see?' Dad called down.

'It's a long passage with neatly packed walls, Dad,' I replied.

Below ground it was always warmer than at the surface, and even though the space had been vented by the hole, it still smelled damp and stale. The ceiling was low and scarred with chisel marks. They created eerie shapes in the light of the lamp. I walked up the passage and found an old clay pot full of water. I wondered who had last touched it. I continued up the passage, which ended in a wall of tightly compacted stone and mud. The passageway had been closed off by the miners who had worked there hundreds of years ago.

I walked back a further fifty feet past the ceiling entrance hole until I reached a pile of rocks and soil that had fallen from the roof and blocked off the passageway. Cavers knew these as boulder chokes.

I reported back up to the surface.

'It's a section of mine, about a hundred feet long, closed at one end with a rockfall at the other,' I called up.

'Be careful,' Dad called down. 'Stay away from the roof fall. We will be with you soon.'

The environment was familiar. I had been in many similar underground tunnels and chambers, and I knew that if you were lucky, you might find relics from the past dropped by the people who worked in these places centuries ago, such as old horseshoes and broken clay pipes.

The team on the surface carried on chiselling away at the small hole I had dropped through and after about an hour Dad, who always told someone where we were and left notes under the windscreen wiper in case something happened, dropped through, followed by Mum, and then the others. We explored the tunnel looking for potential areas we could start a dig through to break into other sections of the mine. Graham and Mick started to take measurements.

However, with school beckoning in the morning, it was time to go back 'upstairs'. We surfaced into the moonlight at midnight. Mum lit the camping-gas stove and put on a kettle to warm us up with a cup of tea. We excitedly discussed returning at the weekend to start a dig through the boulder choke at the end of the passage. However, our plans never came to fruition.

A few days later a large, mysterious, funnel-shaped hole appeared in the same field. The very mine that I had been climbing in days previously had completely collapsed and tons of earth and rock had filled it in, leaving a large sinkhole at the surface. The collapse was covered by the local newspaper, *The Surrey Mirror*, with commentary from Dad. I've still got the press cuttings to this day.

Dad and I were acutely aware of how lucky we were to have escaped being buried alive, thinking we must have a guardian angel looking after us.

*

I was born on 23 July 1962 – a tumultuous year. The Swinging Sixties were just getting going, the Soviet Union and the USA went head-to-head in a potentially apocalyptic game of chicken during the Cuban Missile Crisis, and the space race was hotting up. On 22 July, NASA launched, then aborted, its first inter-planetary probe, Mariner 1. The following day, on the other side of the Atlantic, Mum had a more successful mission when, after a long labour, I was born. My parent's first and only child. The subject of why I never had a brother or sister didn't come up through the years and, to be honest, it wasn't ever something I thought about because our family life was so idyllic. The three of us did so much together that it never felt like anything was missing. Mum and Dad were happy as one-child parents and I was happy being an only child. I had good friends and I was never lonely. Together, we were so adventurous and busy that another child might have held us back.

Mum had a tough start in life. Originally from County Carlow in Southern Ireland, her mum died of tuberculosis when she was eleven. Her father, Ben Carroll, a plasterer and builder, was unable to work and look after her and her sisters at the same time, so he put them into a convent where they were raised by nuns, while he moved to the UK to find work and a home where they could eventually join him.

My paternal grandmother, Edna (her nickname was Dolly) was a lovely woman and we had a very special bond. As a child she suffered a terrible accident when a scalding-hot pan of fat tipped over her head, leaving her permanently scarred and almost bald.

Dad was born in 1930, and after he left school he did an apprenticeship at a local printing company, The Surrey Fine Arts, as a stereotyper – which involved casting the printing plates out of molten lead and engraving them before they were used in the print presses. In 1952 he was called up for national service for two years and went into the elite Airborne Artillery. Dad completed the gruelling training and was awarded the coveted maroon beret, which he was very proud of, and completed his military parachute course at RAF Abingdon.

Mum used to do a bit of housekeeping.

My parents met at the Oakley youth club in Merstham when Mum was eighteen and Dad was twenty-four. They made a good-looking couple. Dad was five feet seven inches tall, slim and very fit. He was always smartly dressed and looked after himself. Mum was a gorgeous, gutsy four-foot-eleven-inch Irish girl with dark hair and radiant blue eyes. They were both adventurous and loved the outdoors, and they passed that on to me.

From the off, I had an unstoppable adventurous streak and could often be found clambering onto the roof of the row of garages at the side of the flats. In those days kids were free to play and have harmless fun. Parents would let their kids roam and explore without panicking. Mum would keep a loose eye on me from the window of the flat, but generally she would let me get on with it as the area was considered safe.

As I became more confident, I would run as fast as I could across the roofs of the garages and launch myself into an adjacent tree. There were other young families living in the same block and the kids all boisterously played outside the flats and on the garage roofs like a pack of wild monkeys. I can distinctly remember the caterwauling of one elderly resident living on the top floor who constantly yelled at us out of the window to get down off the roof and get out of the trees. Every so often there was an accident, like the time one of the kids decided to run across the roofs of a row of parked cars dressed in a Batman costume and went straight through the fabric sunroof of a Morris Minor.

I went to the newly built Spring Vale Primary School and family life had an easy rhythm. Dad worked Monday to Friday; he made a modest living but we didn't have a lot. He came home each week with his pay packet and would lay out £10 to save for a holiday and put aside some money for petrol. We had an Austin A30 that he was constantly tinkering with in his garage with his mate Wally. Dad was a good mechanic and kept that car going for years. At the weekend he'd fix whatever was wrong with it, buying parts from the motor accessory shop or the breakers' yard.

Mum and Dad always wanted a house, and when I was five we moved to a place called Woodhatch in Reigate, Surrey. The property was a rundown two-bed, semi-detached former council house that cost £3,000. It was fair to say it was a 'doer-upper'. Dad was a grafter and handy at DIY, so he was unfazed by the work required.

Dad was intent on passing his practical knowledge on to me. He had antique muskets in the house and together we made gunpowder which we then loaded into the old pistols and fired into the local stream, creating an impressive bang and puff of smoke. If I did that today, I'd be arrested under the Terrorism Act.

Bonfire Night was always a big deal in my house. We made fireworks with the homemade gunpowder and Dad knew how to create the perfect bonfire and Guy. He sewed and stitched the head using an old sack, then stuffed it with exploding crow-scarers, which he'd buy from the local agricultural supplier.

Normally they burned down a slow fuse and exploded every fifteen minutes, but when packed in a ball inside the Guy's body they went up spectacularly, usually blowing the Guy to pieces.

I got my first blank-firing pistol when I was five and I had my first air rifle when I was seven. I used two-pence pieces placed upright in a bucket of sand at the bottom of the garden as target practice, and at seven I was a crack shot, always walking away with prizes at the local fairground. Most of my friends had air rifles, too, and we'd take them over the fields and put up bits of wood as targets. No one batted an eyelid. Not even the police, who stopped me and my friend Richard once when we were older as we were crossing the road to the fields carrying our rifles.

'Off to do a bit of target practice, lads?' the officer asked.

'Yes sir,' we said.

'Have a good day then. And be careful.'

The police were part of the community, and they knew who to keep an eye on. Likewise, they knew who behaved and they understood that just because you were carrying an air rifle it didn't mean you were up to no good or out to harm someone. Crime rates were low and people respected the law. When crimes were committed, detection rates were good because people knew their local beat officer and trusted them. Ours was John Buchanan, he was always on foot, and he had eyes and ears everywhere. He was also an instructor at the Church Lads Brigade.

My childhood above ground was perfect but it got very interesting at a young age thanks to a chance encounter that my dad had.

Dad was an action man, and despite working five days a week he managed to pack lots of other things into his life, one of which was caving, of course. He'd been interested in it for much of his life and as far back as I can remember there were bits and pieces of equipment in the house and shed – such as brightly polished brass miners' lamps and lengths of rope.

Dad's interest became an obsession when one of the reporters on the newspaper where he worked was covering a story about some old abandoned mines in the area that had been rediscovered by a man named Dennis Musto, who later helped lower me into the hole in the Surrey field. The reporter knew Dad was a caver and invited him along on the assignment.

Dad duly dusted off the old carbide caving lamp, went along to the mine entrance that Dennis had found and met another caver there named Robin Walls.

The area had been a centre for mines and quarries as far back as medieval times, the earliest of which was recorded in the Domesday Book of 1086. The main material mined there was Reigate Stone, which was a pale-green to light-grey stone that was particularly prized because it was easy to carve. It was once one of the most important stones used for building in London, and in particular in the construction of Royal palaces, including parts of the Tower of London, in the Great Hall doorway at Hampton Court Palace, in Windsor Castle, Westminster Abbey and in Henry VIII's Nonsuch Palace.

Over the centuries the mines, long since abandoned and forgotten, had collapsed and then been filled in before being reclaimed by the earth. The entrances and shafts had become overgrown, and there was no reliable record of where they were, or how far the network reached.

Dennis was determined to rediscover the underground world, and after years of research he had started to search the local woods around Merstham. He looked for signs in the ground, depressions or piles of old firestone that might give him clues. In the corner of a wood at the bottom of a large embankment he found a possible entrance and started digging a vertical shaft. After several weeks he broke through a section of tunnel, which became known as Bedlams Bank Entrance. Dad and Dennis met right at the beginning of this underground odyssey when the newspaper article was being written, and Dad's sense of adventure and imagination was sparked like one of his crow-scarers going off.

'Perhaps I can give you a hand sometimes?' he offered when they met.

Dennis was more than happy to welcome another pair of hands. Neither man had any inkling at that stage just what a huge task they were committing to.

From then on, every Sunday Dad met Dennis and some other volunteer cavers in the woods and they'd all disappear into a hole in the ground and spend the next few hours picking away at rock falls, inch by inch.

Dad was enthused by the adventure and, after about four weeks of exploration, suggested 'bringing his little nipper along'. That was me.

'How old is he?' Dennis asked.

'Five,' Dad answered.

'Five?' Dennis frowned.

'Yes, he's good. He's small and can help us in the tiny tunnels,' Dad insisted.

The following week I was presented with my first little helmet rigged up with a miner's light and, like something from a Dickens novel, I was taken into the woods and, tied to a safety line, I climbed down the shaft on a flexible wire caving ladder. Once in, I took stock of my surroundings. My senses were firing all sorts of information at me. Through the beam from my headtorch I could make out the shape of the tunnel and could see that the floor was layered with small boulders, the air around me damp and musty. It was completely silent and any movement I made sounded loud and echoed off the walls. Dad blew out the flame on my carbine lamp along with his and we were in total darkness. The darkness and the silence would have been enough to scare off most adults forever, but for me it was the most exciting and mesmerizing moment of my short life up to that point. The adrenaline rush was instant and I was hooked. I wanted to be a part of this adventure and couldn't wait to go back each week. I was proud to be involved. I wasn't scared or apprehensive.

Good progress was made, and several sections of the tunnel had been opened by the time I started helping. Some sections were high enough to stand up in and walk through, others required a crouch on hands and knees. We came across stalactites and evidence of previous human activity, such as symbols scratched in the walls, footprints in the ground and bits of mining equipment left behind. There were cart tracks grooved into the floor where the miners had pushed trolleys full of rock or spoil that needed to go to the surface to be disposed of. Sometimes the tracks seemed to stop dead at a roof collapse, as if the cart had magically passed through solid rock. When we reached these dead ends, we knew they were collapsed sections. This is where more digging began, slow and steady through compacted rock with a hammer and chisel. Sometimes the passageway was so low and narrow that we would be on our bellies. Dad was up front at the rock face and I was behind him. We called these sections 'crawls'. He pushed the loosened rocks and debris underneath him and back towards me with his feet and then I did the same, passing the rubble underneath me, then back with my feet down the tiny tunnel to one of the other guys, who would load it onto a sledge that we had built and drag it out to empty in the main passage. It was like a production line.

It was always exciting when we broke through the collapsed wall to the other side. The air quality in the new space was tested each time before anyone went in. This was done by reaching through the gap with a candle or a carbide lamp and if the flame reduced, it meant the oxygen level was low and the space needed venting before we could enter. When the gap was too small for one of the adults, I was sent through. We always broke through at the top of a pile, so I'd crawl through the hole and scramble down four or five feet on the other side into a new passage, which was completely dark except for the light on my helmet. Mine were the first pair of eyes that had seen those reopened sections for hundreds of years. I stood there in the quiet, looking around, taking it all in. In those moments, it was all mine. I was there on my own, a child deep in a mine. It was a very special feeling. The passages varied in length and could be hundreds of metres long. Tentatively, I'd walk or crawl along the new passageways into the uncharted space while the adults waited for me to report back what I'd found. Some of the sections we opened were 800 feet under the surface and whenever we broke through into a new, significant piece of mine we wrote our names on the wall neatly to record the moment as part of history.

That became the pattern of our weekends for many years throughout my childhood and teenage years. We went down the mines at ten in the morning and we'd come up four hours later, blinking in the sunlight. Time went quickly underground.

As we ventured deeper and further, one of the other cavers, Graham, mapped the system and we stashed caches of chocolate, water and equipment such as matches and candles in waterproof bags as we went, in case we ever got lost or trapped.

My caving gear included boots, jeans, a top, an old jacket, gloves and a helmet. I was still too small to fit into any overalls or knee pads but Mum sewed canvas knee patches onto my jeans. We used to carry a shovel each and we packed water and spare candles in a canvas Second World War gas-mask bag. The calcium carbide for our lamps was kept in a metal ammunition tin. The flames in the lamps were fuelled by a highly flammable gas called acetylene. It was produced by water dripping from the top chamber of the brass lamp onto the small pieces of gravel-like calcium carbide in the chamber below. As the water landed on the chemical it produced gas, which was pushed through a felt pad and emerged in a fine jet in the front of a reflector. The jet was ignited by a small flint wheel. The acetylene light produced an extremely narrow, bright white flame. You could adjust the brightness by controlling how much water dripped down. The fuel lasted around four hours and we took plenty extra. Battery powered miners' lamps were available, but my dad always said, 'once your battery goes, son, then you're buggered.' Another benefit of the carbide lamps was that if we ever did get trapped, the bottom of them could be used as a heat source to keep hands warm. Another vital piece of kit was a flexible wire ladder that was wrapped around a bar placed over the top of the shaft and allowed us to climb up and down the shafts.

We must have looked quite a sight, emerging from a seemingly innocuous hole in the ground, covered in mud, in the middle of the Surrey woods. As soon as we were back above ground we took off our filthy jeans and boots, put them in a bag and washed our hands before driving the old Austin A30 home. In the back garden Dad and I would then clean all the equipment with a bucket of warm water. Boots were scrubbed, lamps were washed, dried and polished with Brasso, then clipped back on the helmet bracket and placed on the stairs again, ready for the next week's adventure. This discipline of maintaining my tools and equipment and ensuring they were ready for the next use was instilled in me at an early age and has stuck with me ever since.

After Dad, Dennis and the team opened Bedlams Bank they started looking for another section of the mine to the east called Quarry Hangers. They spent years digging through backfills and eventually broke through into an area they thought was 'The Hangers'. There they found an old oxen skeleton and clay pipes used by the miners. This was a very tidy section of the mine, with ochre troughs and the names of miners and dates chalked on the working faces of the rock. The cart tracks in this section remained intact.

Dad was always extremely diligent when it came to safety and he erred on the side of caution if ever there was even the slightest whiff of peril. However, he also knew that caving and under-ground exploration came with inherent risks and that even with the best mitigation practices in place there was always the chance of something going pear-shaped. And one day it did, when Dad was down a much deeper section of Bedlams Bank.

He was digging through a roof fall in a chamber not far from the oxen skeleton when it suddenly caved in on him. Unconscious, with head and facial injuries and a broken arm, he was frantically dug out from under the debris by Dennis. Luckily Dad was wearing his helmet and he regained consciousness shortly after. It took them about three hours to get back to the entrance shaft where Dad had to climb the swinging wire ladder with only one arm. Not wanting to burden the ambulance service, he then drove himself to East Surrey Hospital on Redhill Common and was admitted for two days. He had to have his arm reset and, frustratingly for him, he was in a cast for several weeks. But as soon as the cast was off and he was fit enough, he went straight back underground.

Dad would also gleefully recount the story of how when he was a child he had been buried alive in a sand cave in a quarry on Redstone Hill in Redhill. He was playing in the cave when the tunnel collapsed on him. It was only the small gap between his legs allowing just enough oxygen to come in that kept him alive until the local firefighters dug him out. Subsequently, Dad's words of wisdom to me in case I was ever in a collapse was 'try to curl into a ball and get your head between your legs; curling up in a ball protects the head and creates an air pocket.'

Dad and Dennis were becoming well known for their achievements in the mine, and one day a large earth mover used in the construction of the M23 and M25 went through the roof of a mine in Rockshaw Road, Merstham. On behalf of the company, French Construction, Dad and Dennis carried out an inspection of the tunnel and advised the contractors where the tunnels were.

A large area of the mine to the west of Bedlams Bank was to be filled in with concrete to prevent the new motorway collapsing when it eventually carried vehicles. Mass volumes of liquid were pumped through bore holes on the surface. Fortunately, prior to the tunnels being filled with concrete we were able to remove some of the large stalagmites for preservation in the local museum.

The mines were my childhood playground, so full of wonder and excitement. As I got older, life was one long fun-filled adventure. School was fine. I liked science and sport; I played football and ran for the cross-country team, representing the school in the county championships that we won, but I lived for the caves and mines on the weekend. Caving and mine exploration became my passion. I enjoyed anything practical but got bored with the academic stuff. I did my real learning away from the classroom and just wanted to play in the garden, go to the nearby woods, put my tent up and spend time down the mines or go to cadets. My enthusiasm was infectious and my fascinated school teachers asked me to give a talk to my classmates about my underground exploits. I took some artefacts that we had found to show off and my geography teacher helped us date the pipes by researching the angle of the heads.

Our caving kit went everywhere with us, even on our annual Cornish camping holidays. We went every year to the same campsite, Presingoll Farm Caravan and Camping site in St Agnes, which was, and still is, run by Pam Williams. In fact, I still take my family there every summer. Dad drove the trusty Austin A30, with our gear and a suitcase perfectly packed and strapped onto the roof rack. The journey took about eight hours chugging along with the 800cc engine that Dad meticulously maintained. One summer, the big end went on the crankshaft as we approached St Agnes and we crawled into Peterville Garage to ask the owner if he could make the repair. He was too busy but offered Dad the use of his tools and the corner of the workshop to do it himself. While I spent a great week surfing in Trevaunance Cove, Dad was up to his elbows rebuilding the car's engine.

We took our caving maps and lots of rope and explored the vast network of the Blue Hill mines, often entering the horizontal passages – known as adits – from the beach. One summer, a huge storm tore through the campsite for two days. Tents were blown to shreds, everything was sodden, caravans got rolled onto their backs and everyone bailed out and went home apart from us. Everything we had was soaking wet, but resourceful Dad got some bin liners and fashioned ponchos for us. We took our clothes to the local launderette, dried all our stuff and spent a day re-establishing our camp. After that, we enjoyed two weeks of solid sunshine in an empty campsite.

At eight, I joined the Cubs, at eleven, I joined the Junior Training Corps, which was part of the Church Lads Brigade, where I learned to play the tenor drum and bugle. At thirteen I joined the air cadets, which I loved and I relished any chance I got to go flying and gliding. We went to RAF Abingdon, where Dad had done his parachute training, and we got to fly in a Chipmunk with an RAF pilot.

The experience began with the young recruits being shown a military safety film on how to bail out of a doomed Chipmunk aircraft and deploy the parachute. In clipped English the narrator explained: 'In the event of an emergency, climb onto the wing, jump, count to three and pull the ripcord.' After the safety briefing, we were given our parachutes, which were strapped to our backsides on a harness and were designed to be used as a cushion in the aircraft. Throughout the entirety of the flight in the two-seater Chipmunk I hoped the engine would fail because I really wanted to jump out of the plane just like they showed us on the film.

I loved the cadets and stayed for three years. While all my mates were at home or hanging around the street at the weekend, I'd be off flying, gliding or caving.

By the time I left school at the age of sixteen, I had an enviable skill set. I was resourceful and resilient and could problem-solve quickly without flapping. All those amazing experiences made me who I am today and the driving force behind it all was Dad, who I loved, admired, and wanted to be like. Thanks to him and Mum, my childhood was like growing up in a real-life version of *Raiders of the Lost Ark*.

Chapter 2

I left school with only a vague idea of what I wanted to do and a strong idea of what I didn't. I knew I wasn't destined for an office job, or a regular nine-to-five. Ideally, I wanted to be outdoors and doing something practical. I also liked the idea of being my own boss.

I had always taken an interest in the emergency services, particularly Fire and Rescue. As a small child I watched in awe as fire engines whizzed past, wondering what dangerous and exciting incidents the men and women on board were about to encounter. I was drawn to the idea of saving people, of being the rescuer. My favourite TV programme was *Thunderbirds*, which followed the exploits of the Tracy family who make up International Rescue, a secret organization founded to save human life. They were helped in their mission by technologically advanced land, sea, air and space vehicles that were called into service when conventional rescue methods proved ineffective. The most important of these vehicles were the five 'Thunderbird machines'. I imagined myself as Jeff Tracy, head of the family and rescue organization, and dreamed of building an identical Thunderbird machine that I could fly and use to rescue people in peril anywhere in the world.

I had no idea that my rescue interests would later shape

themselves into a career. There were no job ads in *The Surrey Mirror* for trainee rescuers, or Thunderbird pilots, but the thought was there, and it remained a seed for the next few years through my late teens and early twenties as I tried my hand at a variety of different jobs.

Like most teenagers, I learned to drive as soon as I was legally allowed, and after eleven lessons with a driving instructor in a Mini Cooper (one of the old ones that really were 'mini'), I passed my test on the first attempt. I had had plenty of practice behind the wheel prior to the test because Dad regularly let me drive with him on L plates. I once drove my parents all the way from Surrey to Cornwall on a provisional licence, avoiding the motorways. My parents saw this period of time as payback, taking full advantage of having a designated driver, and had me chauffeur them on Friday evenings when they went out for a drink.

Our trusty Austin A30 had finally given up the ghost, despite Dad's best attempts at mechanical life support. The replacement family car was a Vauxhall Viva. Dad also had a moped, a little 50cc DT Yamaha, which he preferred to use if there was ever a choice. So, after I passed my driver's test, he handed the keys of the Viva to me and used his moped to commute instead. In return for the much-appreciated gift of my first car, I drove him to work when the weather was inclement. Dad also had a rare 1962 Lotus Elite, his pride and joy, which I still have today.

The freedom of the open road allowed me to begin a side hustle taking groups of people on tours down the mines. It was purely a word-of-mouth venture. I began taking friends and Scout groups and quickly became an unofficial underground tour guide, running evening jaunts through the ancient tunnels for a few extra quid.

Often on a Friday night I met groups of office workers in The Feathers pub in Merstham, not far from the mines, for these clandestine trips. They followed me to a lay-by near one of the access shafts where I explained the safety drill. They changed out of their office garb into more suitable attire, the wire ladder was lowered, lifelines were attached and then we climbed down into the darkness.

There were no authorities regulating the mines and no health and safety regulations. By that time plenty of people were venturing into the shafts themselves with no guide or experience. The tunnels that we had painstakingly spent years opening and mapping had become a magnet for other cavers, thrill-seekers and teenagers looking for somewhere to smoke dope and drink booze. People travelled miles to explore them, and the local caving club eventually 'adopted' them and became quite territorial over access to them. Sadly, I even noticed that our names were scrubbed out on the breakthrough plaques on the wall, in what I assumed was a petty attempt to hide the fact that Dennis Musto, Mick Clarke, Graham and my dad had opened up the Bedlams Bank mine. Luckily, we had the photos for the history books. There were also other excellent cavers to mention who came along with us, such as Stuart Goldsmith, a firefighter, and Ron Smith. They were two regular guys, cavers who were always down the mine. Ron worked in a boarding school and would often take groups of Scouts down the mine with him. There was a small community of cavers who we'd regularly bump into.

One day, Dad and I met members of the Fire Service up the mines who were carrying out training in the tunnels to familiarize themselves with the environment. They didn't know their way around so they only ventured a short distance, using numbered white gardening pegs with orange fluorescent tape around the top as markers. I offered to take them through the mine to show them how extensive, disorientating and dangerous the tunnels were, and they quickly realized how vast the system was and how out of their depth they were, should an incident occur.

After further discussions with the senior officer on the job, I suggested that perhaps they should keep our numbers on file and call us should they need any training, or someone to lead them, a scout. He agreed and over the following weeks I held some training sessions for the Service. Soon after, I was given a pager connected to Surrey Fire and Rescue control centre, which was used to alert me if there was an incident in the mines that Dad or I could help with. I felt honoured that my expertise and knowledge was valued by the emergency services and I diligently and proudly carried the pager around with me. This was my first 'official' role as a rescuer and I'd be lying if I said that the idea of the pager going off and me racing to the scene of a rescue didn't give me a thrill.

The first time it beeped was late at night. It was on my bedside cabinet where I left it each night. I was asleep but I woke with a jolt of adrenaline. I'd been versed on the protocol. If I was available, I was to call the Fire Brigade control room, tell them who I was and I would be given details of the incident.

I dialled the number.

'Control room?'

'This is Peter Faulding,' I said.

'We have an incident at Merstham, a number of children are reported missing. They entered by number eight entrance,' the control answered.

'I'm on my way.'

Since being given the pager, I carried my caving equipment in the back of the Viva, so I didn't have to waste time collecting it if I was called. I dressed quickly, ran downstairs, jumped in the car, gunned the one-litre engine and sped off to Merstham. Even before I approached the rendezvous point that control had given me, I could see the blue lights illuminating the low cloud. I realized something serious was happening. When I pulled into the road there were six fire engines, five ambulances, police everywhere and a group of very worried-looking parents. I parked, got out and started pulling on my caving outfit.

A police officer came over.

'Who are you?' he asked.

'Peter Faulding,' I replied. 'Special rescue.'

'Follow me,' he said.

I was taken to see the crew commander, who I knew from the training sessions I had carried out for the Special Rescue Unit for Surrey Fire Brigade.

'We've got a group of Scouts lost down there. They went down at around 3 p.m. so they've been there for nine hours. No one reported them missing until late. We don't know if there are any injuries and we don't know what section of the tunnels they are in. We need to find them urgently.'

I nodded and with that I was escorted to the mine entrance and went in, followed by a team of six firefighters carrying a collapsible stretcher and medical equipment.

I had walked the length and breadth of the system hundreds of times and knew every twist, turn and chamber. I tried to work out what sort of route they would have taken. As you entered the mine it split into two tunnels, the east and west. Most groups ended up going to the east because it was a much larger section of mine. I could also see lots of footprints in the mud where people had gone in. I took the right fork to the east of the mine that led to a honeycomb of tunnels that went for miles. The tunnels were about five foot high in places, less in others. Miners were shorter back when the system was operational but any average-sized person would have to stoop, even where the tunnels were at their highest.

I walked with my head leaned to the side. I could move quickly through the passages, but for tall people it was a struggle. Some of the lowest (which were called crawls, for obvious reasons) were only about twelve inches high and you had to crawl through these on your belly. We started searching every small chamber and over the top of every roof collapse calling out as we went, then stopping and listening for a reply. Nothing. We ventured further in and repeated. At points we left candles to mark the areas we had already searched and also as beacons for the lost party to follow, if they were moving through the system.

Eventually, when we were several hundred metres in, I heard a faint reply. The acoustics in the system were difficult to inter-pret and to orientate around, but I understood where they were coming from, and I headed in that direction. A few minutes later I entered a small, cramped chamber in which sat a group of seven terrified but relieved Boy Scouts with their Scout leader, who took one look at me and the firefighters behind me and started sobbing. They had been sitting in complete darkness without food or water for what must have seemed an eternity, the batteries of their torches long-since dead and their supplies finished hours ago.

I led the whole party back the way we had come and within forty minutes we were all safely back on the surface where several of the parents who had been waiting anxiously ran to hug their children.

I went over and spoke to the Scout leader, who looked dazed. 'Are you okay?' I asked.

'There were monsters in there,' he shivered. 'I saw them coming out of the walls.'

Years of experience taught me that fear does strange things to the mind in desperate situations and in the pitch darkness his imagination had run wild, creating hallucinations. When I took groups down the mines I would get them to turn their lights out, to experience what pitch black feels like. It is so dark you can't see your hand in front of your face even if you are wearing white gloves. In that atmosphere sounds become amplified and even the drip of water from a stalactite can be terrifying.

The success of my first rescue meant I was called upon more as a volunteer to assist Surrey Fire Brigade. Although I had no formal training, my knowledge of the mine system, combined with years of underground crawling through tunnels, meant my experience was valued by the firefighters.

Although the majority of the cavers worked together for the common good of the caves and those enjoying them, there were a couple of individuals who became territorial over them and would openly criticize my efforts to assist the Fire Service. They called me Surrey's 'tame pet rescuer' in the group's newsletter. On occasion someone would try to stop me going down the mines because they thought I would need rescuing, and I won't lie, in those moments I did drop in my father's name and explained to them that he opened most of these mines. That usually did the trick.

I was undeterred and determined to follow my goal to become the new Jeff Tracy.

My role as scout and adviser eventually led me down paths I could never have imagined.

But back then, in my late teens, I needed a job, so I started looking around for ways to earn a living.

My experiences in the air cadets, along with Dad's glowing recollections of his time in the Airborne Artillery, made me consider a career in the Armed Forces. I wanted to follow in Dad's footsteps and join an elite specialist unit. The Parachute Regiment, colloquially known as the Paras, is Britain's elite airborne infantry and I had been holding on to the recruitment papers for this unit handed out by army recruiters before I left school. Parental consent to register for selection was required for anyone under eighteen, and when I placed the forms in front of Dad, he shook his head.

'You need to get a trade, son,' he counselled.

I was disappointed at the time, but in hindsight, along with the 'head-between-the-legs' tip, it was the best guidance he ever gave me.

Instead of trooping off to join the Armed Forces, I put my practical skills to use in another way and applied to be an apprentice engineer.

I had not been academic at school and was told once that I'd 'never get anywhere and would end up on the scrapheap.' My application was rejected due to a lack of the required qualifications, but that just made me more determined. Staying positive,

I persevered, and kept applying for other jobs, eventually securing an apprenticeship with a company called Multico that made woodworking machinery. It paid £19 a week, as cash in a brown envelope with a pay slip. I had a Honda CB50 moped (which I've still got) that my parents lent me the money for. I rode to work in all weathers.

I worked at Multico through 1978 when the Conservative government of the time introduced the three-day working week to conserve electricity, which was severely restricted owing to industrial action by coal miners and railway workers. The country was in political turmoil, but it did give me an opportunity to spend more time exploring the mines and working on my car and the moped.

It came as a shock when, less than two years into my employment, I was suddenly made redundant. Luckily my training officer, who had been assigned to me from the Engineering Industry Training Board, found another opportunity for me. He arranged an interview with the Civil Aviation Authority (CAA) at Gatwick Airport for an apprenticeship as a radar engineer at the Telecommunications Engineering Establishment. It was a dream job at the time, so I put my suit on, polished my shoes, went to the interview, and was offered the position.

At the CAA, I started working in the instrument-making section, where I learned to make intricate pieces of radar equipment from precision drawings. I spent time in the drawing office and in the workshop making components. I also went on site visits to airports to install radar systems. It was a great job and I gained masses of knowledge and experience, but I always felt an urge to strive for more and eventually I got itchy feet and left. My decision to leave was made after one of my ex-Royal

Marine mentors called Les retired and invited us to celebrate. One of the guys in our group, Barry, said: 'Let's go and get hammered on the old fart.' I had an epiphany. I didn't want to be Mr Average retiring after forty years of hard graft only for people who hardly knew me to use it as an excuse to get drunk. I finished my six-year apprenticeship and all my college exams and moved on. I was now a highly qualified engineer.

Away from work, I continued to help as a volunteer rescuer for the Fire Brigade and also advised the service informally on underground rescue. My network of contacts within the Service was growing, and I happily shared advice and guidance on confined-space rescue.

Thanks to my parents, my core values were hard work, decency, respect and honesty. But I also spent my late teens and early twenties enjoying life. I had a good circle of friends and we got up to the usual japes that young people did. I went to the pub, I went to discos, I enjoyed a beer, and I was always game for a laugh and a practical joke.

I also had several hobbies that were unusual for the time. At the age of sixteen I had made my first parachute jump. People have studied the psychological effects of parachute jumping and have concluded all kinds of benefits. It trains the mind to act faster under pressure and teaches individuals to conquer fear, thereby increasing confidence. It encourages calm thinking and self-control. Others say that people take up jumping out of aeroplanes as a hobby because the stress response becomes addictive, giving rise to the classic adrenaline junkie who thrives on danger. None of these reasons led me to take a parachute course and my first jump; I did it because Mum and Dad arranged it. It was a typically idiosyncratic Faulding family outing – a weekend parachute course at RSA Parachute Club, Thruxton, which was run by an ex-member of the Paras, Bob Acraman.

Nothing can prepare you for your first parachute jump. It happens quickly and the adrenaline surge is unlike anything you can imagine. Even though your brain knows all the right safety systems are in place and the risks are small if you follow the techniques you've learned, your instincts are screaming at you: 'THIS IS NOT NORMAL!' The skill of jumping is making sure your logical mind overrides your instincts, which tell you to stay inside the plane.

I loved it. It takes a few seconds from leaving the plane for the canopy to open completely and in those seconds your life is literally hanging by a thread. When you hear the reassuring sound of the air filling the parachute above there is relief and then the sheer joy of drifting to the ground with the most magnificent views you'll ever witness. There is also an overwhelming sense of being alone and being in control.

I was hooked.

Luckily there were second-hand ex-military bail-out packs on sale at the centre. These were long parachutes designed for fighter pilots should they need to bail out of their planes. They had been converted to be suitable for the kind of jumping I was doing. I was treated to a canopy and in the following months took up jumping as a hobby. I later got my own jumpsuit and did half a dozen more jumps with a view to working towards freefall jumps, an even more challenging and adrenaline-fuelled type of parachuting.

To qualify as a freefall skydiver, you need to complete a certain number of successful static-line jumps before moving on to 'dummy pulls', during which you practise pulling a ripcord and deploying your own parachute while still on a static line.

I had reached this stage in my training when I was taken up in an unfamiliar aircraft, a de Havilland Dove – much bigger than the Cessna 180 – which was kitted out with an unfamiliar system. I was unsure and voiced my concerns before we took off.

'Don't worry, you'll be fine,' the instructor said.

I remonstrated.

'But I've never exited from this aircraft before.'

'Don't worry about it. You'll be fine.'

I had a bad feeling, nonetheless.

At 2,500 feet the group began to shuffle forward and one by one exited the plane. When my turn came I turned to the instructor for a last-minute recap on instructions but instead he pushed me out the door. I rolled out sideways, rather than spread-eagled. The canopy was on my back, folded into an envelope-type sleeve that opened and which, because of the angle I was falling, wrapped itself around me. I was upside-down plummeting to the ground, wrapped in a large piece of material and unable to deploy my parachute. Mum and Dad were watching on the ground, horrified. I started to spiral, and my short life flashed in front of my eyes.

'I'm going to die,' was my first thought. But then, in what felt like minutes but would have been nanoseconds, I clicked into survival mode and my brain spilled out information. Get rid of the sleeve. Deploy the reserve.

I did each in quick succession. The reserve didn't have springs to ping it out of its case so I had to throw it clear of me manually, which meant pulling the front-mounted ripcord. As the elastics opened the bag containing the reserve, I held it in place with one hand and reached the other hand in behind the packed reserve to throw it out. By the time I did this I was around 1,000 feet from the ground. The reserve deployed, but by that point the first canopy, free of its sleeve, had also opened. Suddenly I had two parachutes above me. It was a mess and ungainly, but I got to the ground safely. Thankfully the only damage was to my underpants. But the near-death experience put me off parachuting and my canopy went up on top of my wardrobe after that and stayed there.

I also took up scuba diving and during the summer would often go to the South Coast with a friend to dive. It was a hobby I enjoyed, and I did it for several years. I went every few weeks and had all the kit. To me, it seemed like a normal thing to do, but in hindsight parachuting and sub-aqua diving were not the normal pursuits of a teenager. As life progressed, and I started work, I found I had less time to continue these adventurous hobbies and the diving was put on the back-burner until later in life.

One evening, when I was seventeen, I met Mandy at a disco in Dorking and we became an item. She would later be my wife. However, I was still on my quest for a career and a higher calling.

Chapter 3

Driving through Croydon one day, I spotted a big sign outside some barracks which read: *10th Volunteer Battalion: The Parachute Regiment.*

I wrote down the details. Given my lucky escape years before, perhaps I should have driven past. I can only surmise that maybe I was an adrenaline junkie after all.

The selection process started in Finchley, North London, where 3 Company were based. I trained every weekend and every Tuesday at Mitcham Road. On the weekends I drove up to Finchley on Friday night and stayed overnight in a sleeping bag on the drill hall floor.

At 5 a.m. on a Saturday morning in November, 184 of us started the process. The first test was a five-mile run to weed out any shirkers. It soon became evident who had done some training prior to selection. One big guy took to the parade ground full of confidence in a woolly hat and T-shirt with cut-off sleeves. He was jumping around like Rocky Balboa, ready for action, steam billowing out of his mouth.

'Piece of piss this, lads,' he sniffed.

We got split into smaller groups and were required to keep pace with the instructor assigned to us. Our instructor was Geoff Butler, who served with 2 Para. After three miles the instructors started picking up speed. Rocky was in my group and at the 3.5-mile mark he keeled over and was out of the process, along with many others who were dropping like flies.

The following weekend those who had survived the first weekend of physical and mental torture were issued with their webbing kit (belt pouches), military clothing, sleeping bag, boots and a numbered 7.62 self-loading rifle from the armoury. We had to put sand in plastic bags then seal them into packages with gaffer tape. The packages were then put into our webbing pouches to simulate equipment. The webbing was weighed by the instructors and had to be thirty pounds. From the morning of issue, your rifle and thirty pounds of webbing would go everywhere with you, including on runs.

In addition to the physical challenges there were practical lessons. I learned how to strip and fire a range of weapons, self-loading rifles, machine guns, anti-tank weapons, was taught about fieldcraft and camouflage and how to make webbing silent so you could launch ambushes in the dark.

As the course progressed, we ventured further afield to the Brecon Beacons to carry out long battle marches and section attacks with live ammunition. We went to St Martin's Plain military camp in Kent for two weeks. There, we were put in a chamber full of tear gas wearing a gas mask and full nuclear, biological and chemical warfare suits. Inside the chamber we had to remove our masks and say our name, rank and number – it was impossible not to breathe in the tear gas. If you got it wrong, you were kept in until you got it right. Outside people were gasping for breath, coughing violently with eyes streaming. Some were throwing up.

The runs got harder and the kit got heavier. Regular 'tabs', as the Paras called the runs, required us to carry a general-purpose machine gun over ten miles in group formation. The gun stayed with the front runner and after several hundred feet he would hand it to the person behind then drop to the back of the squad to recover, before rotating through the field to the front again in a repetitive circle of hell. We learned to cook rations and sleep under temporary shelters called 'bashers'. Most of our patrols were done under cover of darkness, when we learned how to move silently and stealthily over any terrain in order to strike a deadly blow.

All this intense training was working you up to the fearsome P Company (P Coy) Pre-Parachute selection week – the ultimate test that every Para needs to pass to get into the famed regiment.

P Coy selection was a week of pure pain, involving even harder runs, assault courses, trainasium (an aerial confidence course), log-carrying race, stretcher race, steeplechase, assault course and 'milling', which was basically beating the living daylights out of your opponent.

P Coy selection didn't disappoint. It began with a ten-mile timed battle march (running and fast marching) over undulating terrain carrying thirty pounds of webbing and a bergen, plus two water bottles and a weapon. It was a roasting-hot day and it was difficult to run while pulling your water bottle from your pouch, having a sip and putting it away. Two more recruits succumbed to the pressure on the ten-miler.

Next was the trainasium – the aerial course comprising platforms, ropes, nets, balance obstacles and climbing frames high in the air with no safety nets. It was designed to test a candidate's ability to overcome fear. One of the sections involved running along raised planks fifteen feet above the ground and jumping across a gap. One recruit hesitated slightly before the jump, lost momentum and fell straight through the scaffolding. He was stretchered off the course and failed.

On a similar section you were required to run across the aerial course, jump onto a seesaw designed to throw you off balance, then continue running before leaping over a huge gap into a vertical net on the other side. As you flew towards it you needed to punch through the air and aim for a gap in the net and hang on. If you got it wrong, you fell fifteen feet. On the final section, sixty feet up, you walked across the parallel shuffle bars, bent down, touched your toes, recited your name, rank and number, then stepped over two blocks on the bars so you were balancing on one leg.

Then came the stretcher race, in which teams carried metal stretchers weighing 175 pounds over five miles of undulating terrain. This is done while wearing thirty pounds of webbing and carrying a nine-pound rifle. It was gruelling and my body felt like it was about to fall apart.

We finished with something called milling; it was described as 'controlled aggression', but there was nothing controlled about it. We beat the hell out of each other. If you bottled out or you couldn't fight, you didn't pass P Coy. I was up against a guy called Jock from Glasgow who had no front teeth. We went at each other like gladiators. When the bell finally went, our faces were bruised and battered with bloodied noses. The Para physical training instructor (PTI) held both our gloved hands into the air. It was an equal bout. In those days we didn't wear head protection or mouth guards – no one ever mentioned brain damage.

It was a tough, solid week of challenges that I'll never forget.

At the end of it, just twenty-seven of the original cohort had passed, including myself. We finally got rid of our floppy-peaked camouflaged hats and were presented with the Parachute Regiment's coveted maroon beret with silver winged parachute emblem, topped with the Queen's crown. There was no fanfare, which I felt was fitting for an elite regiment.

P Coy allowed me to embark on the next stage of training, which I was most looking forward to. To be a paratrooper, you need to jump out of aeroplanes. Parachute training was conducted at RAF Brize Norton under the guidance of the RAF at No. 1 Parachute Training School. Their motto was 'Knowledge Dispels Fear'. It was a much more civilized affair than P Coy selection; recruits were driven around in coaches and the accommodation was better. Brize Norton even had its own on-site nightclub, The Spotlight Club. And there was no need to pummel each other in the gym. I imagined these perks were designed to balance the fact that you got thrown out of aeroplanes as part of the course.

I loved parachute training. At that stage in my life I was still working at the CAA, which, as part of the Ministry of Defence, gave me as much reasonable time off as was required for military training.

My first proper jump on the course was at RAF Hullavington, in Oxfordshire. Paratroopers deploying into battle generally make static-line jumps, so that was the focus of the initial training. The first jump was out of a tethered barrage balloon which was raised and lowered on a winch attached to the back of a lorry. The maximum jump height was 800 feet above the ground, which is not very high.

On the day of the first aircraft jump, I walked out onto the tarmac at Brize Norton with my fellow recruits and looked awestruck at the Hercules sitting there, waiting for us. Close up these machines were monstrous, like fat beasts, crouched low on the ground. Somewhere in the back of my mind, a ten-year-old version of me was excitedly exclaiming: 'You're going to fly in Thunderbird Two!'

There were fifty of us in the group that boarded the craft, striding purposely up the tail ramp. I felt like I was walking into a movie scene, and the excitement and anticipation in the belly of the beast as the door slowly closed was palpable. The plane shook as the huge turbines started, and the surroundings rumbled and shuddered as we roared down the runway and took off.

The drop zone was a place called Weston-on-the-Green, just up the M40. There are no windows to look out of when you are sitting on the webbing seats, just a few small portholes above eye level nearer the top of the fuselage. We were jumping without equipment, apart from our helmets and two parachutes – the main one on the back and the reserve at the front. This is called jumping in 'clean fatigues'. We went in groups of eight, one after the other from the same side door, in a process known as single sticks.

We were directed by the loadmaster in the middle of the cabin who controlled everything. When ordered, we stood up and clipped the release strap from the rear parachute onto the over-head line that ran through the cabin. The loadmaster ushered my group forward with a thumbs-up signal. The parachute jump instructor (PJI) opened the door and kicked out the step. The atmosphere in the cabin changed in an instant. It was like suddenly winding down the window in a speeding car and poking your head out. The noise of the engines and the air rushing in was deafening. The PJI dispatcher gave the command: 'Action Stations'.

We each shuffled forward with one hand across our reserves and the other holding the strap that secured us and our static line to the cable. I was in line, standing in the door of this giant metal bird with the wind rushing around us and everyone else shuffled up tight behind me.

I watched the countryside pass 800 feet below. As the pilots approached the drop zone (DZ) of Weston they communicated with the loadmaster through his headset. He gave the thumbs-up to the PJI dispatcher and the DZ red light pinged on, meaning we were over the drop zone.

'Red on!' the PJI yelled.

My heart raced. The red light turned to green. That was the signal to jump.

'Green on. Go!'

The PJI knocked my arm down and whacked me on the back of the parachute.

I launched myself out of the door. My body was thrown around in the slipstream of the mighty Hercules and seconds later the canopy opened. I looked up to check the chute had deployed correctly and scanned around me to see the other recruits floating down to the ground. I prepared for the impact and, with a thud, executed a perfect landing. It was all over in less than a minute and before I knew it, I was bundling my canopy up in a large green bag and trudging across a field to the buses waiting to take us back to base.

The field of recruits continued to diminish. One of our number wouldn't get on the plane, another had refused to jump when the door opened.

The training was brilliant and over the course of two weeks jumping techniques were drummed into me to the point where I could do them with my eyes shut. We trained every day and the course rolled past in a blur of training, jumping, landing and jumping again.

Eventually, those of us who passed were lined up outside the aircraft hangars at Brize Norton and the senior PJI walked along the line, shook our hands and presented us with our wings, which were worn proudly on the right arm. There were big celebrations in The Spotlight Club that evening.

Being a member of an elite unit meant that in addition to the regular Tuesday night drill sessions, I was also away for up to three weekends a month, usually somewhere wet and inhospitable. There was generally lots of physical exercise involved. The requirement to stay fit was paramount.

Life as a Para reservist was exciting and exhilarating, but also demanding. I used to come home from exercises shattered. My shoulders were worn through with bruises, my heels caked in blisters and my feet in agony. I'd go home to Mum, lie down on the sofa and sleep so heavily she couldn't wake me, so she just left me there. Then I had to go to work on Monday morning.

I also went for selection for the 21 SAS and passed the initial selection runs, but the commitment was too much and would have affected my future. My relationship with Mandy was also becoming more serious. In all I was in the Paras for six years and loved it, but as I started working I found I had less and less time and something had to give, so eventually my time in the military came to an end.

Still, I felt I'd done something. I had completed some of the toughest military training and had served with the Parachute Regiment, with an immensely loyal band of men who would die for each other.

Chapter 4

I was living two lives in my early twenties; there was the humdrum working week, trying to earn a living, then there were the weekends, full of action and adventure, either on manoeuvres with the Paras or underground, exploring tunnels and crawling through mines. My heart was in the mines and tunnels, and I knew that there must be more opportunities to use the experience of over two decades in confined spaces. I was still first to be called by the Surrey Fire Brigade when they needed assistance underground, but I needed to work out how to turn the hobby I loved into a career.

While I was working at a company called Air Call, I was introduced to a man called Keith Pillinger, who lived in Bristol and ran his own private ambulance business, Bristol Paramedic Ambulance Services. (In 1998 he made headlines by developing what was thought to be the world's smallest ambulance after converting a Reliant Kitten.) His was the first private ambulance company set up in the South West of England covering accident and emergency calls for the then two private hospitals in Bristol. It grew to be the largest private ambulance company in the country, serving many private and NHS hospitals.

Keith's business model showed me that there was a market for private emergency response services and that the private sector could support the emergency services and work effectively along-side them. His vehicles were authorized to use blue lights and his drivers had been specially trained to do so.

Keith was an interesting man with a strong West Country accent; we became friends and occasionally I rode along as an observer in his vehicles, as I was eager to learn about emergency response. Thanks to Keith, I started to think seriously about how I might turn my unusual skills to better use.

I knew there was a demand for confined-space training across the emergency services – not just firefighters but paramedics, too – because whenever I was called out to the mines to rescue lost cavers, I was always asked questions by the other responders. I enjoyed sharing my knowledge and confined-space rescue was and still is a specialist field. Unsurprisingly, few people had the type of expertise and experience I could offer.

One of the turning points of my life occurred when Dad and I started to do training for the special rescue unit of Surrey Fire and Rescue in the mines that we'd explored. During one training session, four members of the London Fire Brigade rope rescue team joined, one of whom was Pete Gwilliam, a larger-than-life character at six foot four inches, with whom I would eventually work. We took them through the mines to the far sections where the old oxen skeleton remained and taught them how to carry out search patterns and tunnelling.

In 1985 my beloved nan, Dolly, passed away. She'd had a lump in her groin for years that never gave her any problems, and it was assumed to be a benign cyst. Her doctor eventually suggested she should have it removed, and she agreed and went into hospital for the minor procedure. When they sent a sample for analysis, however, it was discovered to be cancerous. And it started to spread rapidly. It was devastating to watch her go from a fit, healthy woman to a shadow of herself. She was so brave, and she kept upbeat and positive throughout her illness. Eventually Nan was in and out of a hospice, something she wasn't keen on. I visited her most days and she seemed to be doing fine. One night she was in her bed, watching TV, chatting away, when I said goodnight and went home. The next day we were told she was in a coma and only had four days left to live. It was heartbreaking to see her like that; she was being pumped full of morphine and drifting in and out of consciousness. I asked the nurse why she was unconscious as she had been fine when I left her the previous night. I was told she had been in a bit of pain during the night so was given morphine to keep her comfortable.

'When will she wake up?' I asked.

'We are just keeping her comfortable,' the nurse replied.

I didn't realize then, but she wouldn't wake up.

I learned later of a process called 'indirect euthanasia' where doses of painkillers are administered to relieve suffering but eventually become lethal. If I'd known at the time, I wouldn't have let them do it, because she seemed happy and was walking around a few days earlier. The end came so suddenly. Just as the nurse predicted, four days later she died, as all her organs failed.

I was in pieces and her loss affected me for a long time. I still miss her and visit her grave often. She was a big part of my life.

It was the first time I'd lost a loved one and it made me realize the importance of dignity in death and how painful grief is. Me and my parents wanted closure and we wanted answers. That experience of loss taught me empathy for others who were grieving.

It was a time of much change in my personal circumstances, too. A few years later, in 1988, I married Mandy, when I was twenty-six. We bought a small flat in Redhill and then in November that year our first daughter, Natasha, was born and we moved to a bigger, three-bedroom terraced house in Horley.

It was around this time that I began to struggle financially.

In 1991 inflation was running at over 8 per cent and interest rates started to rise. In 1992 my second daughter, Danielle, was born and soon after interest rates spiked again. Mortgage repayments became crippling, rising from a few hundred pounds to over £1,000 a month in the space of less than a year. It was impossible for me to meet my financial obligations, especially with two little mouths to feed. The mortgage company, Abbey National, wanted to foreclose and I went to court to try to work out a financial solution that would allow me to keep the house. Eventually a repossession order was granted, and I was informed that my young family and I had just one day to move out. Bailiffs turned up the next morning to change the locks. We were given just enough time to collect a few possessions.

We stayed with my parents in their two-bedroom house, but a short time after we were lucky enough to get a council house in Knighton Road, Redhill. I didn't want to rely on the council, though, I wanted to fend for myself.

While all this personal drama was happening the training I was delivering for the Surrey and London firefighters expanded to include the United Kingdom Search and Rescue Team (now called Urban Search and Rescue Team) and the Helicopter Emergency Medical Services (HEMS). I didn't get paid for the training – I did it on a voluntary basis, claiming just minimal expenses. I taught topics including safety procedures and equipment, the dos and don'ts of getting into confined spaces, extraction, and shoring techniques for building collapse work. Each course was tailored according to the specific roles of the attendees. I taught them many of the important lessons that I'd learned over years in the mines.

One such very important lesson was never to go into a confined space without testing the oxygen levels in the air first. Many deaths occur because people are unaware of the invisible dangers in enclosed environments. For example, each year several seamen and ship workers die when they enter the rusty holds deep in the bottom of a ship, because rust eats up the oxygen. Grain silos are also dangerous because grain absorbs oxygen, leaving the enclosed space unable to support human life. Dangerous gases such as hydrogen sulphide and methane can also collect in low-lying and enclosed spaces, such as manholes, sewers and underground vaults. Methane released in the decomposition of organic material is also a hazard, as it is highly flammable.

I never actively promoted myself, but word quickly spread about my specialist knowledge. As I became more established and trusted as an expert in the field, my details were put on file with the HEMS desk at the London Ambulance Service control centre in the same way they were with Surrey Fire Brigade. Any calls that came through to the control room that required helicopter assistance were put through to the HEMS desk, and if an incident required specialist confined-space assistance, I was called.

I also started flying with the HEMS teams every few weeks as an observer. One day while returning to base after an incident, I was sitting in the back of the helicopter with the doctor and paramedic, observing the two pilots upfront in their flight suits, helmets and mouth speakers, awestruck and thinking 'these guys are so cool'. Having finished school with no relevant qualifications, I felt a pang of envy and regret and thought that I would never get to fly such a beautiful machine.

I was on call to assist, not just for underground and confined-space rescue, but also for incidents that required a climber or someone with technical expertise in rope work. The London Fire Brigade had recently disbanded their own rope rescue team for cost-cutting purposes, having just spent hundreds of thousands of pounds on new light rescue vehicles and a substantial amount of rope rescue equipment, which was subsequently given away to community groups. It was a pattern I witnessed with depressing regularity over the years; specialist teams within the emergency services were shut down to save money because the frequency of specialist calls were few and far between. It was only when a major incident occurred, and the lack of specialist resources was exposed, that the issue was highlighted. Then there would be a public inquiry and it all started again. A new unit would be formed with a different name at added cost. It became a vicious circle. Everyone blamed the government, but the decisions were made at a local authority level.

In those days London was (and still is) a busy city full of emergencies, and so the calls I received from both the Fire Service and HEMS became regular.

One of the first was an incident on top of a crane in Kingston, in October 1994, on a blustery autumn day. A disturbed young lady had climbed 150 feet to the top and was threatening to jump. A HEMS doctor in a response car had been called to the scene because it was too windy for the helicopter to land. Fire crews, an ambulance and police officers were also there.

I raced to the scene after being called by HEMS control. The area and road underneath the crane had been cordoned off and traffic was building up in the surrounding area. Police were anxious to resolve the situation. Firefighters at the scene were helpless as they had no rope rescue capability, and they could not get an aerial platform into the middle of the building site. I parked as near as I could and went to introduce myself to the police, who were controlling the incident. The senior officer briefed me that the lady's boyfriend had committed suicide after developing AIDS and that she was extremely agitated and had been threatening to jump from the crane. She had been throwing spanners down the internal well to keep anyone from climbing up to get her and had threatened the crane operator, who had locked himself inside his cab.

'She's causing a public nuisance, so we need someone to go up and get her down,' the officer in charge of the incident told me.

'It's not as simple as that,' I explained. 'She's clearly in a distressed state, I can't just go and grab her and drag her down. That would be incredibly irresponsible and dangerous. I need to talk to her and persuade her to let me bring her down.'

'Whatever it takes, she needs to come down,' said the officer.

I went to get the kit from my car. I harnessed up, put my helmet on and got rope and another harness for her. My plan was to chat to her, calm her down, make her comfortable, secure her in a harness and lower her down safely.

I trudged over to the base of the crane and started the long climb to the top. About a third of the way up, from where I could be heard, I called up to the lady.

'My name is Pete, I'm a rescuer with the Air Ambulance Service and I'm just coming up to talk to you. Is that okay with you?'

The answer came seconds later. Clang. Clang. Clang. A spanner spun past my head. I carried on climbing, calling out to reassure her as I progressed. She must have realized I wasn't a threat as no more tools were lobbed in my direction.

As I reached a platform near the top I shouted to the crane driver in the cab to reassure him. The man told me he was fine. The lady, who was only in her early twenties, was perched precariously on safety rails. She was thin and shaking. Her cheeks were tear-stained, her eyes bloodshot and she was only wearing a pair of jeans and a light jacket. She looked desperate, exhausted and wretched. I felt sad for her.

'Go away, I'm going to jump,' she said weakly.

I concluded that if she really did want to end it all, she would have done so already. 'Look, I'm not going to harm you. I just want to talk to you. I'm not the police. I'm here to help you,' I said.

I'd had no training in such matters, so I just asked her about herself and why she was there. We started talking, she told me about her boyfriend and how alone she felt. She said she didn't want to live without him.

'You look hungry,' I said. 'Let me get you some food.' I thought that would be a good way of befriending her.

She nodded. I asked her what she wanted, and she said McDonald's. As luck would have it, from the top of the crane, I could see there was one just a few hundred metres away.

'Okay, I'll go down, get some food, and then we can chat some more,' I said. She agreed. The thought of food seemed to divert her attention away from taking her life.

I climbed down.

The inspector met me at the bottom.

'Can you get her a McDonald's? She wants a cheeseburger and a coffee.'

'Are you having a laugh?' The officer was incredulous. 'All she's getting is arrested.'

Thankfully, the doctor was more understanding and went to get the food, which I put in my backpack and carried back up to the top.

By the time I reached her, the wind had picked up and was whistling loudly through the steel tubing of the crane tower. I sat down again and handed her the food and drink.

'Have this, and then I think we need to get you down,' I said gently.

She became agitated.

'I'm not going down. They're going to arrest me,' she said defiantly.

I tried to reassure her.

'No, they won't. There is a doctor waiting for you. He'll take you to hospital and you'll get help. You have my word. Let me go down, speak to them and make sure you'll be taken care of.'

She agreed and once again I climbed down to talk to the police.

'She's getting nicked,' he sniffed when I explained the situation.

'She needs help,' I implored, 'she doesn't deserve to be arrested.'

The doctor backed me up.

'She is traumatized and mentally unstable, I will escort her to the hospital,' he asserted. But this was the nineties and people were not as enlightened about mental health as they are today.

In the stalemate that followed between law enforcement and medical care, I climbed back up the crane and told the lady that the doctor down below was waiting to help her. I gave her my word again and hoped that the cop would see reason when confronted by this poor unfortunate person who was no threat to anyone. She finally agreed and I carefully put her in the harness and lowered her down the middle of the crane, clipped to me.

At the bottom I unhooked her, and we walked out of the structure together. To my horror three police officers grabbed her, threw her to the ground and handcuffed her.

'What are you doing?!' I yelled.

She screamed and squirmed.

'You lied.'

'Let her go,' I shouted.

To his credit, the doctor ran over and started screaming at the police too.

'Get off this girl, she needs help. This is a medical matter; you are not authorized to do this.'

The police refused to admit their behaviour was unreasonable but did decide to let the lady go. The doctor and I helped her up, wrapped her in a blanket and took her to the nearby ambulance.

I was left standing there, in shock at what I'd just witnessed. I was appalled. The firefighters who were at the scene couldn't believe it either and stared at the three police officers, shaking their heads.

At this juncture, it's worth stating that I deal with police all the time in my job and 98 per cent of the officers I've worked with and alongside have been brilliant. They are a force for good and do a great job, protecting the public and solving crime. But in all walks of life there are a few bad apples. I call it the 'two per cent club', and on that job I met one of the two per cent.

Up until that point, I had had the rose-tinted view that in emergency incidents all the different services worked together for the common good of the person in distress or in need of rescue. While that thankfully remains the case in most incidents, disasters and crime scenes I've attended, issues such as crossed purposes, territoriality, ego and stubbornness do occasionally muddy the waters.

I discovered this during another incident not long after. Some children had been reported missing in Poulter Park, in South-West London, having been seen crawling into a storm drain. It was the height of summer, and the drainage system was too tight for the burly firemen at the scene complete with their breathing apparatus. One of the firefighters who attended one of my training sessions asked his control room to call the HEMS desk to get in touch with me to help. I arrived, was briefed, and duly kitted up. I had a gas detector with me and a chemical rebreather. These clever bits of apparatus fit into a small pack and contain a chemical called potassium super-oxide, which decomposes to form oxygen when in contact with moisture. It gives around 130 minutes of chemically generated oxygen. The only disadvantage is that it blows very hot, but if the alternative is suffocation, it is worth the discomfort of a burning throat.

The drainage system I crawled into was wet and very cold as I started my search, with the firefighters providing backup on the surface. As I went along, the multi-gas detector beeped every few seconds to tell me it was working. I spent about forty minutes inside the storm drain searching for the kids, without any luck.

When I emerged a good while later, another crew had since taken over and one of them walked over to me.

'Who the hell are you?' he said.

'I've been searching in the drains,' I explained.

'Bloody members of the public,' he tutted. 'You anoraks go down the drains, get in trouble and we end up having to pull you out.'

I was confused.

'Your team called me to help,' I remonstrated.

'We don't call civilians in,' he said.

He was getting increasingly agitated, so I decided the best thing to do was just to walk away. The kids, I learned later, had found their own way out in another part of the park hours ago, even before I turned up, which added insult to injury.

Later that day I received several calls from London firefighters I knew who told me that a tele-printer message was sent to all stations warning of a member of the public who was pretending to be a rescuer. It named me.

I complained directly to the chief fire officer and a retraction and apology was accordingly sent out to confirm that I was indeed qualified and authorized to advise and assist in specialist rescue.

The HEMS team particularly valued my services and put me on a trauma life support course, which was a two-week training that doctors went on to learn emergency procedures. The course meant that I was confidently able to help the Air Ambulance medics, handing them kit when they asked for it and managing patients onto the helicopter.

I continued volunteering and observing with the HEMS teams for several years and the range of incidents I was called to assist with increased.

One night I was woken by a call from the Air Ambulance

control desk alerting me to a job at Heathrow Airport where my assistance was required.

'There is a truck in a hole with a person trapped,' the operator advised.

The accident had happened on the vast Terminal Four construction site, where the driver of a dumper truck had lost control of his vehicle and driven through the crash barriers around the rim of a huge hundred-foot-deep access shaft. The sides of the shaft gave way under the weight of the truck and the driver and his vehicle slid in. Luckily there was reinforced steel lattice within the hole that broke the fall and acted as a kind of giant springy trampoline. The injured driver was trapped in his cab, in an upside-down lorry at the bottom of the giant shaft.

It sounds like a cliché to say the scene resembled something from a disaster movie, but that's the best way I can describe it. When I arrived, there were fire crews and a BASICS doctor (British Association for Immediate Care).

I climbed down long wooden ladders to the bottom of the enormous shaft and was greeted by the mangled mess of a giant earthmover. I was amazed that anyone had survived. The man inside was screaming in pain and was hanging upside down, held in by his seatbelt. He looked like a broken puppet, and had sustained a nasty head injury. A firefighter was by the side of the wreckage with his head inside the cab; there was very little room for anyone to get in and get the man out, but he needed urgent medical care.

The doctor was there too and was reassuring the man.

'Pete, can you get in to see if there are any other signs of injury while I sort out sedation?' he asked.

The firefighter at the scene was a big guy and too big to get in. I was smaller and experienced at working in confined spaces so was the natural choice to go in. I nodded and squeezed through the narrow gap into the cab and noticed a big puddle of red diesel.

'If this goes up, we are all going to be cooked alive,' I thought.

The driver was confused and delirious and as I squeezed in, I tried to reassure him and check him over. He started to lash out.

'It's okay,' I said. 'We're here to get you out.'

He was a big guy and continued to thrash around and then, without warning, I took a hard punch to the face that stunned me. Thankfully the doctor managed to squeeze in too and quickly and calmly sedated the man who, within a minute, was like a rag doll. I managed to cut his seatbelt and with the assistance of the firefighter carefully lowered and manoeuvred him out of the cab. He was assessed and carried to a waiting crane bucket, then lifted clear of the giant earthmover and whisked off to hospital by ambulance.

At each job I was learning how to deal with people in distress. I now understood that it was important to establish a dialogue with them and make them realize that I was there to help them and get them to safety. Often they were confused and in pain. Tragically, sometimes there was nothing anyone could do to help. I saw a lot of deaths when I flew as an observer on the Air Ambulance and accompanied medics in the response cars.

Interest in my work started to spread beyond the emergency services. One of my underground training sessions with HEMS doctors and paramedics was featured on an episode of a show called *How Do They Do That?* hosted by Eamonn Holmes in the early 1990s. I was invited to the studio to meet Eamonn, who described me as the 'Human Mole'. That has stuck with me ever since. I was featured in the *Sunday Express* and was branded the 'saviour on a skateboard' when I developed a technique for extracting casualties from pipes using a skateboard. One of the Air Ambulance doctors I trained, Heather Clark, wrote a book in which she described me as someone who 'could get into any hole'.

As word spread, I was contacted by HM Customs and Excise and asked to provide training for their confined-space rummage teams who search ships.

By then Mandy, me and the girls were living in a two-bedroomed flat near my old secondary school, but I was still desperate for money at this time, so in addition to the day job I was doing, selling security and CCTV systems, I also signed up to do cleaning shifts at Gatwick Airport at the weekends.

I was then approached by the editor of *Fire & Rescue Magazine* and *Industrial Fire Journal* to write articles for confined-space and collapsed-structure rescue. I put a desk in the corner of the living room and hired a computer and printer from Radio Rentals, as I didn't have the money to buy a new one. I purchased a fax card and software that turned the computer into a fax machine. This became my office. I got paid for writing articles that had a global readership. I had given up loads of time helping out on HEMS and training first responders for free and thoroughly enjoyed it, but this was the first time I'd earned money from rescue work, in a roundabout way, and although money was never a motivating factor, it did make me realize that perhaps there was a career to be made from rescue work.

But the real turning point came when one of my articles was

read by someone at Elf Oil, who contacted me about providing training at the company's refinery in Milford Haven, in Wales. I took a day off my paid job and drove there for a meeting with the safety team to discuss what they wanted.

I realized that 'Big' Pete Gwilliam, who I knew from the training sessions I hosted for the fire service, would be able to help, as he was an experienced fire brigade trainer. He was also good at planning and designing training packages. We spoke and he agreed to come on board with me. We worked well together as our skill sets complemented each other. Elf Oil was my first paid contract. From there we got jobs with other oil companies and found ourselves providing training and support for Shell, Gulf Oil and BP.

In 1995, I finally felt ready to take the monumental leap of striking out on my own. My childhood ambition of founding my own rescue company and becoming the next Jeff Tracy became a reality when I registered the name Specialist Rescue International (later to become Specialist Group International).

It was a big step. I was still earning just enough to get by, but despite having a good reputation, work was not guaranteed. It was hand to mouth, but I enjoyed it immensely. I was helping people and finally felt I was doing what I was meant to be doing.

When I established my company, I went to my local bank manager and requested a £1,500 loan for some equipment. He was dismissive and told me that he didn't think my business plan stood up to scrutiny. He offered me £1,200 instead, for which I was grateful.

It was late 1995. I didn't have time to watch much TV or read the newspapers so I was unaware of events developing sixty miles away, in the countryside outside a town called Newbury, in

Berkshire. In tunnels under fields and in trees in woodland a New Age army was digging in and preparing to fight for a cause that few people understood. Mud-caked, dreadlocked and fervent with a new kind of ideology, these people were about to upset the old order and wreak havoc across the country. And I was going to become their nemesis.

Our paths were on a collision course, and within weeks of the bank manager's vote of no confidence I was to receive a call that would change everything for me.

Chapter 5

In life you can plan meticulously, and generally planning is a good thing, but sometimes you just need that little bit of luck. You have to meet the right person, or be in the right place at the right time. Events were conspiring around me in which I had no input but that would ensure my life took a profound turn. My years spent in tunnels with Dad, the expertise I'd gained and the contacts I'd made in the emergency services all ensured that at the right time I was the only person in the country on the radar of the emergency services who knew how to deal with the growing problem of road protest tunnels.

It was early December 1995 and I was at home when the work phone rang. Mandy answered.

'Hello. Specialist Rescue International.'

There was a pause.

'I'll just get him,' she said.

She covered the mouthpiece and turned to me.

'Mervyn Edwards?' she questioned.

Mandy passed the phone.

'Hello, Peter Faulding speaking.'

Mervyn was from Thames Valley Police.

'A colleague of mine met you at a presentation you did on confined spaces at Netley, in Hampshire, a few weeks ago,' he explained. 'We've got a bit of an issue in Berkshire, near Newbury. I'm not sure if you've seen the media about the Newbury bypass?'

I hadn't.

He continued.

'We've got protesters in tunnels holding up a major road construction project. They're well dug in and extremely organized and I believe you are the man who can get them out. Would you be able to come to a meeting tomorrow where we can tell you more?'

I was intrigued.

That night I pressed my trousers and polished my shoes in preparation. I drove my Vauxhall Cavalier to the meeting the next morning, which was being held in an office at Chieveley Services on the M4, where the construction and Highways Agency teams for the project were based.

Mervyn met me outside. He was dressed in his police uniform and had pips on his shoulders to denote his rank of superintendent, and greeted me with a big smile and a firm handshake. He explained that there were some 'big hitters' in the meeting. They included Treasury solicitors, the Health and Safety Executive, structural engineers, the under-sheriff and the silver commander of the operation, as well as Ian Blair of Thames Valley Police, who went on to be the commissioner of the Met, now Sir Ian Blair.

'Don't worry, they're just normal people,' Mervyn said. 'Just be yourself and answer their questions.'

I followed him into a small conference room. There was a large table in the middle around which sat several men.

'Good afternoon, gentlemen,' I said.

One of the men, a bespectacled, serious-looking chap, got up and introduced himself. He was Nick Blandy, the under-sheriff of Berkshire. (I learned later that an under-sheriff is responsible for the execution of writs in his or her jurisdiction.)

The officials in the room were there to plan an operation that would be the culmination of a controversy that stretched back into the 1980s and centred around the construction of a nine-mile extension of the A34 in Berkshire. Traffic planners said the new road would ease a notorious bottleneck in the town of Newbury, but local campaigners had been opposing the road for years because its construction involved felling 10,000 trees and tarmacking over 120 acres of woodland.

In 1988 there was a public inquiry. Work was due to begin in 1994 but was delayed. In July 1995 Transport Minister Brian Mawhinney made the final decision to approve the bypass, much to the consternation of environmental groups such as Friends of the Earth.

The bitterness between the sides had festered over the years.

In 1992 a protest against construction of a new route on the M3 at Twyford Down added to the animosity between the authorities and the environmental lobby. That protest turned nasty when security guards dragged people off the site. Protesters complained of being beaten up and some successfully sued for wrongful arrest. Many of the Twyford Down veterans were at the Newbury protest, which brought together national organizations, local residents and a rag-tag band of direct-action activists.

They started to move in after the announcement of commencement of work in July 1995. Many took to the trees, living in tree houses – called twigloos – in a practice known as tree-sitting. The first camp was set up at Snelsmore Common. In September

1995 another encampment was established by the Kennet and Avon Canal. A month later another was set up in a copse that was due to be bulldozed. By the time I was called, three further camps had been built and several of the protesters were claiming squatters' rights.

The situation was becoming volatile on both sides; Nick Blandy had received death threats, and a protester, his pregnant partner and their young son who were living in a bus on one of the camps woke to see a man in a balaclava lighting a petrol bomb outside. The protester managed to drive away before any harm could be done and later a local car dealer's son was cleared of attempted arson after telling the court he had drunk ten pints and staged the firebombing as a prank.

Nick explained that police intelligence suggested that one of the methods the protesters were using was tunnelling. They hoped that by digging underground passageways they could obstruct any work as heavy machinery could not drive over a tunnel in case there was a protester in it and the tunnel collapsed.

The presence of the protesters was disrupting the construction work and each day of delay added hundreds of thousands of pounds to the project bill. Both sides were also aware of a European law that forbade interference with trees after nesting season began in early April, so when I was called in December, time was of the essence. A writ had been issued to evict the protesters from the camps, which bailiffs were on hand to enforce, and a specialist team of climbers had been engaged to get the protesters out of the trees. Anticipating public order disturbances, Thames Valley and Hampshire constabularies launched Operation Prospect.

At the meeting I was introduced as the man who could get

the protesters out of the tunnels safely and deal with any subterranean issues.

Mervyn, as silver commander and leader of the operation, then briefed the room with an intelligence picture on what they knew. I took detailed notes.

'When is the operation starting?' I asked.

'Next week. We need to act fast,' Nick replied.

After the meeting he passed me his business card and asked for my quote for the work by ten the next morning.

At that point I had no idea how big the operation was going to be and how drastically my life was about to change. I had all the technical and practical skills they needed, but I had no idea about contracts and costs either.

I discussed it with my parents. The job required more than one person and Dad was certainly someone I wanted with me.

'Dad, we've got a job to do,' I said, and explained what it entailed. 'I don't know how much to charge a day though?'

We had equipment in Dad's shed; ropes, harnesses, props for shoring, shovels and other tools. We'd need to hire a 4x4 (the Cavalier wouldn't cope off-road) and I also wanted to take a medic along in case Dad or I got injured. It would be an early start and we'd be working late into the night. Then we needed to account for the value of our combined experience. I came up with the figure of £3,200 a day and decided not to charge mileage in case that put the bill up too high.

I cribbed together some details, drew up a quote, put it all on a fax and sent it over to Nick Blandy's office the following morning. Within an hour it was sent back, signed. I couldn't believe it – £3,200 a day. I thought I'd won the lottery.

Over the following days I had meetings with the Highways

Agency, the Crown solicitor, Royal Berkshire Fire and Rescue Service, Thames Valley Police and the under-sheriff. I rented a second-hand Land Rover Discovery and recruited a doctor I knew from the Air Ambulance on a daily rate.

The operation to clear the camp was scheduled to start early in January 1996, first thing in the morning the day after 'giro day', when unemployment benefit was paid out. The thinking behind this was that many of the activists were unemployed and would have spent their dole money on alcohol and cannabis the day before, so were likely to be groggy the following morning, making them easier to remove.

Dad and I attended a pre-operation briefing at 2 a.m. It was not known if there were protesters down the tunnel, so our role was to thoroughly search it and ensure it was clear of people or booby traps before it could be filled in. A camouflaged police Covert Rural Observation Post team was in place around the top of the entrance to grab any protesters who came out, like a game of whack-a-mole. The previous night they'd managed to pick off one of the tunnel protesters who popped up for a wee in the middle of the night.

Dad, myself and the doctor arrived in the early hours, ready to go in at first light.

As the thin winter sun rose over the woodland, we walked onto the site of the Snelsmore Common camp along with contingents of bailiffs, police officers, lumberjacks and Highways Agency officials. The site was muddy and as our dawn ambush got underway the camp around us exploded into activity as protesters and officials began to clash. I looked around, taking it all in. It seemed there were hundreds of people suddenly running around and climbing up the trees overhead. The air was full of screams as protesters hurled obscenities at the bailiffs. It was like a war zone.

One of the most popular methods protesters used was called 'locking on'. In its simplest form this involved holding on to, or padlocking yourself to, something solid like a fence or a gate. At the more complex end of the spectrum, it involved grabbing or locking on to bars set inside concrete-filled barrels. Typically, a pipe would be passed through holes in the side of an oil drum or through the top. A bar would be placed through the pipe and the barrel would then be filled with concrete. The protester finally locked their hands to the bar with a loop around their wrist attached to a carabiner.

Moving the barrel would potentially break the protester's arm. The only safe way to free them was to cut through the barrel and the concrete very carefully (often the concrete also contained bits of metal or rope to blunt and snarl up the cutting tools). The process of removing a locked-on protester took hours, and sometimes days depending on the complexity of the object they'd locked onto.

As the Newbury operation swung into action, activists around the site headed to their lock-ons and padlocked themselves in, knowing that once they had done so the process of removing them would be slow and costly.

While this chaos was erupting around us, we headed to the tunnel. I passed under a tree in which a team of protesters had set up camp and was immediately soaked by a full bucket of excrement that was poured over me from a tree above. Dad managed to dodge the splash and after the shock subsided, struggled to hold in a laugh. Suddenly the £3,200 fee didn't seem so much.

I couldn't change my clothes, we had to move quickly and there would be no point anyway because where we were heading would no doubt be wet and muddy. And there was no guarantee that by the end of the job plenty more excrement wouldn't be thrown our way.

The ultimate aim of our part of the operation was to clear the tunnels so they could be collapsed by diggers. We needed to check them all thoroughly as it would be unthinkable to collapse one with someone still inside.

We quickly located one tent over an eight-foot shaft shored up with wood. It was professionally done, better than I expected. I knew there would be another entrance and found it under a second tent. It was a shaft lined with three metal oil drums with the bottoms cut out, stacked one on top of the other to create a tube. The bottom drum had twisted and collapsed. Luckily nobody was in it at the time, or they would have been cut in two.

I asked the bailiffs to cordon off the area and instructed that no heavy machinery be used within a hundred-metre radius, as any vibrations could collapse the system below.

Back at the main entrance Dad and I removed the tent and called out to see if anyone was inside. No one answered. I lowered the gas detector to the bottom of the shaft on a line to test the atmosphere for gases or low levels of oxygen. I attached my safety line to the harness I was wearing and descended the shaft while the doctor and Dad stayed on the surface to act as my back-up team. I carried a chemical rebreather on my belt in a tin (called a turtle) which would give me between thirty and a hundred minutes of oxygen if it all went pear-shaped.

The tunnel was about three feet high and well shored. The

walls were boarded with wooden planks. My first thought was that there could be other tunnels hidden behind the boarding. It didn't look like something that had been dug out by an inexperienced eco-activist with a bucket and spade. I later discovered it had been shored up by a Cornish ex-tin miner. The tunnel sloped gently downwards, and as I descended I pulled a line-fed communications set with me so I could speak to Dad at the surface. I crawled on my hands and knees and further down the tunnel I located a small chamber on the left in which there was a large, brown, military ammunition can. I carefully opened the lid and found around a month's worth of supplies, including food, a small, methylated-spirit camping stove and toilet rolls. In a sealed plastic food bag I found a book, *The Tunnels of Cu Chi*. It was about a battle between Viet Cong guerrillas and special teams of US infantrymen called 'Tunnel Rats' during the Vietnam war, which took place in a 200-mile labyrinth of underground tunnels and secret chambers built around Saigon.

I crawled further in, around a corner, not knowing if there was anyone there. I was in my element. Eventually I came to a length of passage half-submerged in cold, muddy water with bits of plywood floating on top. I crawled through the water, sweeping the ground with my gloved hand in case there was anything sharp.

At the end of the passage, I reached the other entrance shaft with the collapsed drum blocking it.

Having made an initial sweep I went back to the surface for a hot drink and to brief Dad about what I'd found and to prepare for the next stage. I needed to cut holes in the panels in the shoring to ensure no one was hidden in a secret chamber behind.

Once warmed up I descended back down the tunnel and started phase two. The conditions were cramped and awkward. By the time we finished it was dark and cold. Luckily, I had my caving wetsuit on under my overalls to keep me warm, otherwise I would have certainly frozen. Evictions were still taking place in the trees, which were lit by floodlights.

Nick Blandy was still there, and I told him the tunnel was now secure, was empty and that he should bring the diggers in case protesters tried to occupy it again.

I was exhausted and Dad and I then had a long drive home, nearly falling asleep at the wheel. The only thing that kept us awake was the smell of my earlier excrement shower, which still clung to my overalls. By the time I got home I'd been awake for nearly twenty-four hours.

The following morning, Nick called and asked me to send over my bill.

'Good job yesterday,' he said.

I sent over the invoice and a short time later Nick called again.

'Peter, I've got an issue with your bill,' he said.

I explained that my quote was included in the contract that I had sent previously and that if he thought it was too much maybe we could look at it again.

He laughed.

'I've got bailiffs charging me forty grand a day. You need to reconsider your prices because you are the man who solved the problem for us. This project is costing millions and you've got the most dangerous job, yet you're only charging £3,200 per day. You haven't put in for a hotel, meals or fuel costs. What hotel did you stay in?'

I told him I didn't, that I drove home.

'What!' he exclaimed. 'No hotels? No expenses? Seriously, consider what you are charging because the way things look now, quite a lot of this type of work will be coming your way.'

It was advice I would become grateful for.

Within a month the number of protesters had increased and there were more than twenty camps along the route of the bypass, with names such as 'Rickety Bridge', 'Quercus Circus', 'Radical Fluff', 'Pixie Village' and 'Heartbreak Hotel'.

The protesters had a lot of support. People didn't like the idea of the British countryside being paved over. In February 1996 around 5,000 people marched along the route in objection to the road. Environmentalists claimed it was the largest single demonstration against road building in Britain. Television presenters Johnny Morris and Maggie Philbin, who lived nearby, joined the march. The skirmishes between protesters and officials made headlines and were regularly filmed and broadcast on national TV news. It was a politically sensitive operation.

The efforts to remove the protesters were sometimes chaotic from a health and safety point of view. There were so many contractors working at so many locations. In one incident they chopped a tree down and it fell onto a contractor. Luckily only the brushwood hit him and he was standing in a peat bog at the time, which absorbed the impact. The Health and Safety Executive (HSE) eventually got involved and carefully governed operations from then on. I had many meetings with Nick Marsh, the principal inspector at the HSE, to review tunnel operations. Tunnels dug by protesters had never been seen before, so we were setting the standard operating procedures for any future operations.

Tunnels were found under several other camps along the Newbury bypass route, and I was called back for other operations – having adjusted my pricing accordingly! One weekend, protesters erected a large tent and held a rave in it. Under cover of the music and mayhem they dug a tunnel, secretly carrying the soil out of the marquee and dumping it in the woods.

The tunnel was discovered the following Monday morning when the revellers had departed. It was hidden under a drain cover. I was called in to clear it. The tunnel went straight down for twenty feet and had been dug into soft sandy earth without shoring or any support. It was a death trap. Me, Dad, Big Pete and a medic were taken in under a cover of riot shields, Roman army-style, because projectiles were being hurled at us from all directions by protesters in the surrounding trees. Thames Valley Police went in hard and arrested the perpetrators. This time I had to carefully shore the tunnel inch by inch as it was extremely unstable. I invited Nick Marsh from the HSE along to see how complex and dangerous the work was.

While I was at Newbury some officers from the National Police College at Bramshill came to talk to me about protests and protest removal. There was genuine concern that this new type of environmental protest, designed to cause delay and disruption to infrastructure projects, was going to become increasingly prevalent as the government unveiled a series of developments.

While many of the projects that they feared would be targeted were on private land, and so legally needed to be dealt with as a civil matter, using writs and bailiffs, Newbury had shown the authorities that numerous public order offences were also occurring. Indeed, over the course of the Newbury protests there were 748 arrests, including two protesters dressed as a pantomime cow (both the front and back end were nicked). Private security cost £25 million and policing cost £5 million. Between Twyford and Newbury, protester techniques and organization had improved and would no doubt evolve further. The cops who came to talk to me wanted to be ahead of this game of cat and mouse that was unfolding across the UK.

On my last job at Newbury, two men in red hats walked over to me as I emerged from the tunnel I'd just cleared. I knew by then that under-sheriffs wore red. One was Nick Blandy, and he introduced the other as Trevor Coleman, the under-sheriff of Devon.

Trevor told me that three similar protest camps had been established on the site of the proposed new A30 near Honiton. There were tunnels at each location, he said.

'There's one particular individual called Swampy who seems to be a specialist at digging these things. We're going to need your help down there as soon as possible because the bailiffs are likely to be going in within the next six weeks,' he explained.

As in Newbury, the protests in Honiton had been bubbling along for several years. The story began in 1983 when the Exeter-to-Honiton improvement scheme was listed in the national road-building programme to enable faster travel through East Devon, avoiding several villages. The scheme went to public inquiry in 1992 following objections and was given the green light by the government in 1993. The flashpoint was a thirteen-mile section of dual carriageway.

During the campaign the protesters staged a rooftop demonstration at the home of Tiverton MP Angela Browning, they invaded Highways Agency offices at Sowton, they took over the offices of contractors WS Atkins and occupied Trevor Coleman's offices, serving him with a fake eviction notice and throwing paint all over his reception area.

By 1996 they had built permanent camps on both sides of a river valley through which the road was due to be built. The camps were named Trollheim and Fairmile. Another nearby camp was known as Allercombe. All included tree houses and networks of underground tunnels and bunkers. Swampy had been in Newbury but our paths never crossed. It was likely the tunnels there were his work. In Honiton he and his tunnelling colleagues had created a network under the planned roadworks, which were thought to be forty feet deep.

I agreed to help at Honiton, and as the year ended, I started planning for my next big protester removal job and a showdown with Swampy, who was hunkered down in a hole in Devon and capturing the public's imagination.

I was always learning on the job. Just as these environmental protests were new to the authorities, they were also new to me. It was obvious that Honiton was a bigger, more complex job because the tunnels, according to the intel, were bigger, deeper and better fortified. I was going to need more equipment.

Whereas in Newbury all the tunnels had been empty, the ones in Honiton were going to be occupied, and that created a whole new set of challenges. As before, safety was the main priority. The aim was to clear the tunnels and remove the protesters without any injuries.

I built my team and picked sensible, cool-headed people, including Andy, a well-respected former SAS soldier, who supplied me with three other ex-SAS men to monitor the tunnels and equipment overnight. I knew these guys would not fall asleep and let me down. I also recruited Dave Barry and my old trusted friend, Chris Edwards, plus a couple of ex-Paras, Sean and Roy, both short, lean guys. Big Pete was my operations manager on the surface.

I purchased extra equipment. I hired air compressors to enable us to send breathing air down the tunnels. I invested in safe lamps that couldn't cause any explosions and a second communication system so anyone 'downstairs' could talk to people 'upstairs'. I bought drilling and cutting equipment, medical equipment, safety harnesses and ropes. And most importantly a tea urn. I spent around £30,000 – most of the money that I had earned at Newbury – but I knew that Honiton was going to be a much bigger and more lucrative job. Finally, I invested in a Jeep Cherokee and a trailer to tow the equipment.

Before the operation began, I attended many planning meetings at Trevor's office in Exeter. The police, the bailiffs, the contractors, the HSE and the Highways Agency all attended. I also met with the intelligence team from Devon and Cornwall Police and was given more intel on the tunnels. A date was set for us to move in, which was kept secret, because if it leaked we would lose the element of surprise.

I was handed a set of sketched plans of where the tunnels were. I studied them carefully and several things about them didn't add up. For one, if the system was as big as the plans indicated, it would probably have collapsed. I decided to keep an open mind about their accuracy.

On the day of the raid all the relevant parties met at a remote industrial estate somewhere in Devon next to the M5. There were around fifty vehicles. The plan was to drive in convoy and raid the smaller camp first before it got light, take over the camp and clear the protesters.

Again, this was ostensibly a job for bailiffs to handle but the police were involved because of the public order element and the potential for criminal damage and injury.

We set off down the motorway ready for a first-light attack. After a few miles we hit a thick fog bank and the convoy got split into two, half wrongly continuing down the M5. There were no sat-navs back then, but luckily we did have mobile phones (the old-fashioned Nokia ones). After many calls between all parties, we arranged to regroup at a point further down the motorway. We lost an hour waiting for everyone to reconvene and plan a new route.

We set off again following the lead bailiff vehicle, turned off the motorway, drove through a small village and up a driveway, which the lead driver thought was the entrance to the protest camp but was in fact the entrance to the farm next to the camp. It was four in the morning by then and the bewildered farmer, awoken by the fifty various vehicles that were driving up his lane came storming out of his house with a shotgun. We got out to explain what was happening only to hear howling, whooping and clanging from the adjacent land where the protesters had heard the commotion and realized what was happening.

By the time we got onto the site some eager bailiffs had irresponsibly already gone down one of the tunnels and roughed up a couple of the protesters, which was totally wrong and against the rules of engagement. It was also stupidly short-sighted. We wanted a cordial relationship with these people. Our job was going to be hard enough without giving them reason to hate us. I climbed into the tunnel and made sure the protesters were okay and uninjured and helped them out.

We started to survey the site to assess what we were dealing with. In the middle of the camp there was a covered shaft. We carefully removed the cover, in case it was rigged up to something, or someone. I looked down a vertical drop of around ten feet to the bottom where I could see a male body, face down, not moving.

I called the doctor over, got the team to set up the tripod over the hole, connected the wire ladder to it and quickly climbed down. The shaft was only about four feet in diameter, so it was hard to get in and check for signs of life without treading on the person at the bottom. I felt for a pulse on his neck and tried to see if he was breathing. He was in a thick coat and his skin felt cold. I got no response. I climbed back out and told the doctor to go down and check him. After a couple of minutes, I heard a high-pitched squeal from the protester.

'What's wrong?' I called.

'I couldn't get a response, so I squeezed his plums,' the doctor said. I'm not sure if this method was in the standard medical handbook, but it was effective.

When the doctor resurfaced a few minutes later he explained that the protester was lying over a lock-on pipe that had been sunk into the ground, his wrist was clipped into a wire noose in the pipe which was half full of groundwater. The protester was soaking wet and cold. After cutting away the sleeve from his jacket, the doctor saw that the circulation to the man's upper arm was cut off, and the limb was turning purple.

'He's probably got half an hour before he loses his arm. You need to get him out now,' I was told.

An ambulance was called in preparation while I climbed back down the shaft and the team lowered some tools to me in a canvas bucket.

I felt around the protester and established what I was up against. In such a tight space, I had no other option than to lie on top of him and dig around him to get into the pipe and free his hand. Even this approach was almost impossible, so I changed tack and began to carve a chamber out of the wall of the tunnel into which I could crouch to get a better reach. The protester was getting colder and increasingly uncomfortable.

I reassured him that we were there to help and that we would get him out safely and as quickly as possible. It was hard, difficult work, but eventually I managed to dig a channel under him through which I could reach into the pipe and cut the wire noose that his hand was in. We were both relieved when he could roll over and he willingly allowed me to fit a rescue harness around him so that he could be winched to the surface and taken off to medics to be checked over. It was a close call and the eviction had only just started.

Through the rest of the day we cleared protesters from other tunnels and completed our objective, ready to move across to the main camp the following day where Swampy was waiting.

The next camp was much bigger and better organized. Once again, we were met by scenes familiar from Newbury. Dirty, muddy and chaotic, with protesters in trees and people being dragged away and arrested.

We established our operation under a canopy, laid out our kit, set up a kitchen area and a medical area and checked all our safety equipment. It was much more complex than just jumping down tunnels and dragging out bodies. The first thing we needed to do was work out how many people were in the tunnels and if they were in any immediate danger. Then we needed to establish air lines and a communication link so we could talk to them and monitor them in case anything went wrong. Once I started working on a tunnel, the duty of care was down to me. Then, as we began the process of getting into the tunnel, we needed to shore it up as we progressed. We had timber on site and tools so we could cut wood to fit. I was the only one who had any timber-shoring experience, so all the pressure and responsibility was on me. Even though the roofs and walls of the tunnels we entered were already supported, it wasn't to a suitable standard to protect the tunnel from collapsing.

We set up near the main entrance of a tunnel the protesters had named Big Momma. We established that there were four people inside, including Swampy (whose real name is Daniel Hooper) and another tunneller named Muppet Dave (he got the nickname because he had a dog named Muppet).

We needed to get a dialogue going with them so they knew that we were not a threat. We needed everything to be calm because any mistakes underground, any violent movements or scuffles, would have dire consequences. The authorities, while anxious to clear the camps as quickly as possible, were also very mindful of the PR implications of any dead or injured protesters. Our game was softly, softly. To a degree, this played into the protesters' hands because they knew once they were in the tunnels they weren't going to be forcibly removed, so the process of getting them out was time-consuming. The longer it went on, the more media attention they got. And in the days before social media, headlines were their goal. They wanted the world to know what they were doing.

I carefully climbed down the entrance shaft into Big Momma and shone my headlamp in. The first thing I saw was a reinforced door covered in corrugated steel and barbed wire with a 'danger' sign hanging from it. I could hear voices and movement behind it and we struck up a brief conversation. I explained that I had an airline and communications equipment for them in case the tunnel collapsed.

'I'm going to leave the airline and comms at the door. If you open the door and pull it through when I go back to the surface, we can carry out a comms check and we will start free-flowing compressed air at low pressure to keep the oxygen levels safe for you,' I explained.

As I retreated, I could hear the bolts being undone, the equipment being pulled into the tunnel and the door being firmly secured again. The compressed air started flowing and we established communications. I was always reassured once those safety measures were in place.

I climbed back up the ladder and was surprised to see a group of bailiffs digging into the ground about fifty yards away, above a section of the tunnel. Concerned, I walked over and asked what they were doing. They explained that they were digging down to intercept the tunnel. I was shocked.

'Do you realize how dangerous that is? You could collapse the whole network. There are people down there,' I exclaimed.

I demanded they stop and explained that we were the only team authorized to work in the tunnels and had carefully planned the operation for weeks. They ignored me and carried on. I immediately called the health and safety inspector, Nigel Chambers, who had signed off all my method statements and risk assessments. Nigel was 'displeased' and drove from Bristol to the site. He walked into the site and instantly stopped the whole eviction by producing his Crown Warrant, which gave him authority to call the shots. An hour later a meeting was called for all parties. The bailiffs were reprimanded like naughty schoolboys, and it was explained that SGI were the only ones who were authorized to deal with the tunnels. He laid down the regulations very clearly, then allowed operations to recommence under close supervision.

Back at work, I started to shore the vertical shaft, taking measurements and shouting them up to Chris on the surface, who was busy cutting the timber to size. Good progress was made but it was slow, hard work. I finished shoring the main shaft at about eight that night.

The following morning, I started to open the tunnel to about three feet square, shoring as I went. This produced a lot of soil that had to be bucketed to the surface with a gin wheel and rope. I was teaching Sean, Roy and Dave how to shore up as we went but I knew it took years of experience to cover all eventualities. About eight feet from the main shaft, I came across the first reinforced door, which was made of wood and again covered in corrugated steel and barbed wire. I managed to cut the door free using a reciprocating saw. As it fell away, it revealed another short length of tunnel and another reinforced door with a gap at the top. Two sets of eyes were peering at me through the gap.

'It's the "Men in Black",' said one of them. This is what the protesters called us, due to the black fire-resistant suits we wore. At that time, I kept my company name a secret from them for fear of retaliation.

'My name's Pete, what's yours?' I said.

'I'm Dan and this is Dave.'

I immediately recognized Dan as 'Swampy'.

'This one is going to take you a while,' Dave said with a laugh. It was all very jovial banter.

'I'm going to have to shore this length of passage before I get to you and then attack the door,' I told them.

'That could be a problem. We've nailed your comms cable across the back of our door. Cut through it and it will cost you,' said Swampy. He was right, the comms cables were manufactured in Canada, and were specially made to be intrinsically safe. They were not cheap and we only had three lengths of a hundred feet each.

'That was a crafty move,' I thought to myself.

Over the following days Swampy and I had several conversations as I worked my way along the tunnel. We spoke every day, many times a day. He was always behind a door or some obstacle, further along the tunnel. We chatted about regular things and it was always good-humoured. I built up trust with him.

On one of my frequent trips upstairs I met a police negotiator who had arrived to try to resolve the standoff quickly.

'We want this job finished,' he said. He held up a large US army-type field telephone set.

'Can you take this down so I can talk to Hooper?' he asked.

Having spent a significant amount of time with Swampy by then, a mutual understanding had developed between us and I knew there was no way he was going to speak to the authorities or come out freely.

'I really don't think he's going to want to talk to you,' I said. 'But I'll ask him.'

Back down I went.

'Dan,' I called, 'the police negotiator upstairs wants to talk to you about coming out. I have a phone for you.'

From behind the rickety locked door, I could hear raucous laughter at this preposterous suggestion.

'Okay, give me the phone,' Swampy said.

He took it and I went back upstairs to tell the officer that Swampy was ready to talk.

He got on the handset and introduced himself.

'Hooper, this is the police negotiator from Devon and Cornwall Police. I'm here to talk with you,' he said in a slow and confident manner.

'Go *$%! yourself,' replied Swampy, and the line went dead abruptly.

'Hmm.' The embarrassed police negotiator was lost for words.

Back down I went again, only to find bits of what was once a police radio, dumped outside the door for me to take back up. I collected up all the pieces and handed them over to the negotiator.

'Probably best if you just let us deal with it,' I said.

We shored up to the door, which I cut out of its frame, taking care not to sever our comms cable. Before I removed it, Swampy, who was small and agile, and Muppet Dave, who was stockier, scurried off out of sight to the left at a right angle into another tunnel. I passed the small door back to Chris, who was in the tunnel behind me, then shone my light into the dark. I could hear muffled voices coming from a small tunnel straight ahead. I crawled in. It led to a T-junction and two cramped tunnels, each containing a protester. Ian was one and Eleanor, aka 'Animal', was the other. Both were in lock-ons. Eleanor's arm was in a pipe sunk into the wall. To release her hand, we needed to carefully shore around her, cut into the pipe and free her locked-on hand without causing injury. One mistake could sever an artery and cause life-threatening injuries.

There is a test to measure the stability of a tunnel which involves putting a wedge in cracks to see if it stays in or falls out. If the wedge falls out over the hours or days, it indicates that the earth is probably moving, the structure is unstable and needs to be evacuated. I tested the tunnel and the wedge stayed in, which meant no immediate danger.

It took several hours to dig Eleanor out, during which she was supplied with hot coffee. Once she was released from her lock-on, she freely crawled out. Ian was amiable, too, and within a few hours he was also removed peacefully. With them out of the way we could continue the search for Swampy and Muppet Dave.

Swampy had gone down a vertical shaft and Dave had gone to a far end of another tunnel and disappeared down a shaft into a chamber with a metal lid on top.

We moved slowly through the tunnel, shoring as we went. The days ground on, the camp was gradually cleared, the trees were emptied of protesters and Dave and Swampy made a last stand. We built up a good rapport with them as we moved forward, and a mutual respect developed between us.

I got Dave out first and before he went to the surface to get arrested we sat underground and had a chat over a cup of tea. These chats were useful as he told me he was going straight to another protest at Manchester Airport afterwards, to dig in where it would be even tougher to evict him.

That left Swampy, who, because of his fame, was making daily national headlines at that point. The eyes of the country were watching and waiting for him to be brought out. Sean and I started work on the vertical twelve-foot shaft that he was in and although we couldn't see him, we knew he was there. The shaft was just twenty-four inches square and had to be widened and shored down vertically. All the while we had to prevent the soil we were digging out from falling down the shaft and entombing Swampy at the bottom. It was a long job and we worked twelve-hour days.

We put a small camera in the ceiling pointing down over the top of Swampy's shaft for safety and noticed that when he was confident the rescue team had left the tunnel for the night he would creep out and come up to the taller shaft wearing just his underpants, where he could move about and stretch his cramped limbs. It was like watching a badger on its nightly prowl.

Each morning we were greeted by a delightful bag of human waste that he had passed out for us to remove. When we finally shored to the bottom of the shaft, I looked into the twelve-inch worm tunnel he was in but couldn't see anything as it was shaped like a banana.

'Are you okay, Dan?' I called.

'Yes, I'm fine, but I could do with a cuppa,' he replied.

I got Sean to bring down a flask, which I pushed out in front of me as I squeezed into the wormhole. Behind me Sean placed his hands against the soles of my feet to act as a platform, which I pushed against to propel me into the tiny hole.

I popped through and there was Dan in his chamber.

'Your time is up,' I said. 'You've made the world's media, cost the project an absolute fortune, and highlighted your campaign to the world. Have a cup of tea and let's get out of here before we both get buried.'

Dan was calm and very reasonable and knew that there was nowhere else he could go. He was tired. He'd been underground for nine days and had survived on water, cold baked beans and Frosties.

'You've done what you came here to do,' I said to him, and we shook hands.

On 30 January 1997, at 8.30 p.m., Swampy came out of the Big Momma tunnel to an explosion of flashing cameras and a sea of media. I was behind him wearing my balaclava. He was arrested and taken away.

Two days before Swampy's last stand Trevor Coleman had introduced me to yet another under-sheriff, Randall Hibbert. Randall told me he had a bigger problem brewing at Manchester Airport on the site of a planned second runway. There were around eighty protesters underground in eleven tunnels. It was where Muppet Dave said he was going next.

Randall wanted them out within three months and asked me to help. It was going to be a big job. In the following weeks I found an industrial unit for sale in Dorking, Surrey, which I purchased and there established Specialist Rescue International's first base. Meanwhile, other protest camps were springing up at the sites of developments around the UK and the protesters were all using tunnels to delay construction. I realized I was going to be very busy.

Chapter 6

They called the Manchester Airport protests the war in the woods, because they were a battle between the airport, which planned to build a second runway, and eco-protesters intent on saving ancient woodland over which the runway was due to be developed.

The trouble started in 1991 when the runway plans were first discussed. The project had an initial costing of £36 million, was due to create 50,000 jobs and was scheduled for completion in 1998. The airport argued that the second runway was required to keep pace with increasing numbers of air travellers.

By 1993 the airport board had chosen the option of building a second runway parallel to the existing one. In 1996 there was a public inquiry into the plans, in which the Manchester Airport Joint Action Group gave evidence along with representatives from Liverpool Airport and former hostage Terry Waite, who grew up in a village that was due to be blighted by the project. The hearing lasted 101 days and on 15 January 1997, Transport Secretary Sir George Young and Environment Secretary John Gummer gave the thumbs up for a second runway. Projected costs had already risen to £172 million and within days of the announcement the first protesters had set up camp.

Into this maelstrom of discontent, in 1997, stepped me and

my team, prepared for an extensive operation in which we were up against very dedicated individuals intent on causing as much delay and havoc as possible. The potential for danger and injury was considerable and the briefings we received from both Greater Manchester and Cheshire police forces (the protest site straddled the border between the divisions) suggested the number and complexity of the tunnels and obstacles we were going to be facing were considerably greater that what we'd found in Newbury and Honiton. It was also likely that there would be the requirement to be working on separate tunnels simultaneously. The protesters had already been there for months and Swampy was also well ensconced by that time, having high-tailed it up there after his stint at Honiton. He appeared on the front cover of a national newspaper wearing a breathing apparatus set that was given to him by an ex-firefighter. He said he was ready to swim through a submerged passage – sump – where no one would be able to remove him. I assembled a team of over a dozen people, including Dad, a medic, and the team I used at Honiton. I acquired yet more equipment and machinery, including a new air compressor, two quad bikes and trailers, which would enable us to get quickly between locations because it was a massive site with areas thick with mud.

Each of these jobs came with ethical considerations. The protesters, after all, were fighting for a cause. I had my opinions on some of these causes. I agreed that roads and infrastructure needed to be built – the road schemes particularly were about trying to ease congestion and pollution – but on a professional level I never let my opinions cloud my judgement. I was there to do a job and to keep people safe.

The eviction started with Greater Manchester Police raiding

the part of the site under their jurisdiction. The bailiffs moved in immediately after and I went over to assess the first tunnel with my team. By this time the protesters knew to leave an air pipe out of the tunnel for us to connect our compressed air pump to, which kept their oxygen levels to around 21 per cent. We established the air supply and set up our compressor fifty metres away to prevent carbon monoxide from the exhaust going into the tunnel.

I carefully removed the wooden lid and took stock of what I could see below. At the end of the tunnel there was a chamber, inside which was a woman. The lower half of her body was in a lower section. Her arm was locked on to a steel tube that had been set into three tyres full of concrete. As I entered, I told her we were going to have to shore up the entrance before we could cut her out of the lock-on. The rest of the day was spent shoring up the main entrance and the first part of the tunnel. She told us that she could unclip herself so she could lie down to sleep, which I assumed she did when we left.

The following morning we returned and she was back in place. It was a cramped passageway so we couldn't use a disc cutter because of the smoke and dust that would have been created in the confined space. Instead, after two hours I managed to separate the top tyre from the bottom two, exposing the steel pipe that her hand was locked into. It took the rest of the day and into the evening to release her.

Each protester and tunnel presented us with a different challenge. The next tunnel we tackled had been dug into the side of a cliff face in a ravine through which the River Bollin ran. It was 120 feet down a marlstone rock face. In it we encountered another young woman locked on inside. It was a tough job as we had to widen the tunnel while hanging from ropes to get her out. Once the lock-on was released, we lowered her down the cliff face to safety.

At the top of the ravine, among the trees where we tied off our ropes for the abseil, we noticed a tent under which we found a hole covered with a steel sheet. The words 'I have a noose around my neck, do not lift' were chalked on it. This immediately set alarm bells ringing.

What happened next illustrated why my careful, methodical approach was so vital. Rather than try to move the cover to peek under it, we first moved the tent and cleared the area of any junk or rocks. I then carefully dug a small trench down the side of the metal big enough to shine a torch into so I could see what was underneath. I spotted a thick, heavy chain welded to the underside of the sheet, hanging down into darkness. We had no keyhole cameras in those days, only an angled search mirror, but it was dark, and I couldn't see past a few inches of tunnel.

We carefully cut the trench deeper without moving the cover until it was big enough for me to crouch in and look down the hole properly with a torch. The shaft went down around ten feet and widened out at the bottom where a girl was lying on the mud floor, face up, with her arms out in a crucifix pose, both arms were inserted up to the shoulders into metal tubes concreted into the walls. The concrete was reinforced with old bow saw blades, pieces of iron and bits of rubber. The other end of the chain that was attached to the cover was padlocked around her neck. There was another tunnel leading from the bottom of the shaft that the lower half of her body was inserted into.

She couldn't move and was chanting 'save the trees'. She had willingly trapped herself and if the tunnel collapsed or someone had quickly lifted the lid attached to her neck, she would have died. It was an incredibly sophisticated and well-thought-out attempt to slow down progress and I couldn't help but admire the extreme lengths these people went to in fighting for their cause.

I gathered my thoughts for a second.

The first task was to cut the chain from underneath the cover so we could move it aside and get in the shaft. But we couldn't simply snip it because it would have dropped onto the girl's face and it was heavy enough to smash her skull. I tied a piece of rope through the chain under the lid and passed it to Big Pete, who secured it. I then carefully cut through the chain with hydraulic cutters. The rope prevented the chain from dropping. The metal cover was then removed further, exposing the difficult and dangerous task ahead.

We set up a quadpod (like a tripod, but with an extra leg), attached a wire ladder to it and I went down to the bottom of the shaft to check on her and set up an airline. She told me that she was one of four protesters in the tunnel system. She also told me that her hands were secured using handcuffs purchased from a sex shop and that she couldn't release herself.

The other protesters were on three different levels below. I knew it was going to take several days to get them out and was concerned that the tunnel would cave in and bury them all alive. I put her into a positive-pressure breathing apparatus mask and a throat microphone was carefully placed around her neck on an elastic strap which allowed her to communicate with the surface. We then built a strong plywood platform over her chest and head to protect her in case the shaft collapsed – this also protected her from falling debris as we set about shoring.

The doctor went down every half an hour to check on her and give her glucose drinks. The other three men deeper in the system were trapped as there was only one way in and out and she was blocking it.

We methodically secured the shaft and widened it at the bottom so we had room to work on her lock-ons. It was difficult enough to cut someone out of a concrete lock-on in the daytime on the surface, let alone at the bottom of the shaft built into a wall in an extremely confined area. With one wrong move an artery could have been severed and she would bleed to death. In total, 'The Worm', as I later learned her nickname was, lay there for three days while we worked around her. Although her welfare was our primary concern and we did all we could to take care of her, there was a limit to what we could practically do. She was cold, distressed, sleep-deprived, and had been lying in her own waste for days. She was taken to hospital as soon as we lifted her out on a stretcher.

We managed to safely remove two of the male protesters in the tunnel the following day. The last protester was down another level and refused to come out. I had placed three wedges in cracks in the ceiling the day before and they had fallen out as the cracks widened. It would have taken days to shore the tunnel properly and there was a real risk that the final protester was going to be buried alive, which he did not understand. I decided to break all the rules and worm my way down the unsupported tunnel to try to talk him out. I found him sitting on a mound in the middle of a round chamber surrounded by a giant puddle of water. It looked like he was on his own island.

'If you don't come out, you'll die. The ground is extremely unstable and moving,' I said.

He acquiesced, grabbed his plastic bag of possessions and we both started the long tight crawl on our bellies to the surface. As he came to the surface he shook my hand and thanked me.

There was no time to celebrate, and work continued in the other tunnels. We had eight compressors running around the clock pumping air, keeping the protesters alive.

In one tunnel we found a girl who was completely covered in mites and lice.

In another I was reunited with Muppet Dave, who was living deep underground in a long tunnel stocked with food and supplies. He'd already been there for some time before we arrived and would go on to spend over twenty-one days underground in the tunnel he called 'Cakehole', which was around 150 feet long with lots of doors that had to be painstakingly removed to get to him. He was living right at the very end of the long tunnel and was determined to stay until the last moment. When we eventually brought him out, he was greeted by Trevor Coleman, Randall Hibbert and Superintendent Alec Barclay from Cheshire Police, who all wanted to meet him because he'd been down there so long. We had a coffee together.

'So, Dave,' Alec said, 'what are you doing next?'

'I'm just going to bum around Greece for two or three months.'

'How will you get there?' I inquired.

'I'll fly out of Manchester,' he joked.

While everyone was still laughing, Dave got up, walked over to the waiting police van, and was arrested.

Once he was out, I concentrated on a lone female protester in a tunnel who could not release herself. Her fingers were starting to go numb. She was in a very narrow tunnel with her arm locked into the floor. I had to lie on top of her to chisel away at the lock-on. We were both hungry and later that evening Big Pete went and got us a Chinese takeaway. I fed her while she lay there, we chatted and eventually I cut her free. On the surface she gave me a big hug and thanked me for looking after her.

It took a whole month to complete the job, during which I hardly saw Mandy or the kids. The job was dangerous, exhausting and stressful, but also very lucrative. So much so that before I left I went to a local car dealership in Wilmslow and put down a deposit on a new soft-top Aston Martin DB7 Vantage Volante, which I had built to order. I'd been underground up to my neck in mud, dirt, lice and human waste for weeks on end, risking my life, and I figured I deserved a treat.

As a postscript to the Manchester Airport story, two years later we were back there again because some of the protesters had studied the planning application and found a small piece of land at the bottom of the proposed runway that wasn't covered by the original writ of eviction. They moved in and dug a huge complex tunnel system into the bottom of a bank, so the entrance was already eighty feet underground. This made it ultra-dangerous. It was another 'Disco Dave special', with several levels and many twists and turns. The entrance was a small, chequered-steel door with a large, yellow smiley face painted on it. It was held in place by a wooden frame with an upturned horseshoe nailed to the top and a postcard pinned to it, which read *Dig for Victory*. As I approached the door to check it I heard a muffled voice behind.

'Is Pete the tunneller around?'

I kneeled and talked into it.

'This is Pete,' I said.

'Please don't drill the door, my back is leaning against it, and I'm locked on.'

We'll have to go under, I thought.

We looked around the area and found their airlines coming out further down the bank. They were attached to a plastic milk bottle and a twelve-volt computer fan, which was not sufficient to supply air to those inside. Big Pete and my dad quickly cut off their rig and connected our line to establish an air supply. Then we tunnelled under the door, leaving the protester unharmed but locked on above us. He was locked on to a very cleverly designed device using a triple layer of two pipes and concrete mixed with rubber. Once we safely removed him, we then came across a protester standing on a ladder in front of us by a vertical shaft with a noose around his neck. He threatened to jump if we went near him. I decided the best way to approach was to dig a forty-five-degree stepped tunnel to intercept the shaft underneath. Then, if he jumped, we could support him from below and cut him free. Each of these challenges took days to complete and were major excavations. We had compressors and lighting running twenty-four hours a day and a three-man night team on duty.

The last protester to be removed on Manchester II was Disco Dave, who had been digging further and further in as we progressed towards him through the weeks, leaving big piles of soil for us to clear. Eventually when I got to him, I knew the structural integrity of the system was failing. I used the wedges again and they dropped out.

'It's going to cave in, Dave,' I said. 'You really do need to come out.'

He trusted what I was telling him and agreed to surface.

We had a camera inside the tunnel that monitored the area that he was in and when we took him back to our command centre for a coffee before he was arrested, we looked at the screen and watched the chamber that Dave and I had been in just ten minutes before completely cave in. My life and that of Dave's would have been snuffed out in an instant had he taken even ten minutes longer to come out. Dave was so shocked at what a close shave he'd had, he collapsed in a panic attack. Our medic immediately set up an intravenous drip and put him on oxygen, and Dave was taken to hospital in an ambulance. Most people do not realize the extreme lengths protesters would go to and the very real life-threatening situations they put themselves and the rescue teams in.

It appeared to be a golden age for environmental protests. The media attention that early protests at Newbury and Honiton received acted as inspiration (and a recruitment tool) for other protests. Prominent protesters such as Swampy were role models for a new generation of activists who rallied to environmental causes. At each protest new and ingenious methods were developed to hamper the authorities, and these were then shared across the wider, loosely affiliated community.

Around that time, I was asked to sit on the Home Office Working Group for Policing of Environmental Protests; a new panel set up to look at best practice and techniques. As I had experience on the ground (and under it), I was invited to share my knowledge with the group, which included senior government officials, Mervyn Edwards, a police commander from the Met and the head of the Health and Safety Executive. My involvement in that group led to many other high-profile protester-removal jobs.

By that time – 1998 – the saga of the Birmingham Northern Relief Road, which became the M6 Toll, had been rumbling on for many years and was reaching a crescendo, into which me and my team would be deployed.

The term eco-warrior had by now become common parlance and the protest camps along the route of the proposed road were manned again by many veterans of the protest movement.

Plans for a new West Midlands motorway were first proposed in 1980 and there had been a public inquiry in 1988 and a tender agreement was signed by Midland Expressway Ltd in 1991. There was another public inquiry in 1994, and as work was nearing commencement in 1997 the eco-warriors established several camps along the route and squatted in numerous derelict cottages that were due to be demolished, which they set about fortifying. One, Boundary Cottage, was fitted with armour-plated doors in readiness for the inevitable showdown with bailiffs when the final legal appeals to stop construction failed.

Sadly, in April 1998, a protester named Dave Richards was found dead in his sleeping bag in Boundary Cottage. He was a fugitive from the French Foreign Legion and his fellow protesters proclaimed him 'the first martyr of the campaign'. Police removed the other protesters from the cottage to conduct inquiries into his death and the property was demolished when they'd finished, which antagonized the protesters, who claimed the authorities had capitalized on Mr Richards's death.

The episode did allow police to get an insight into some of the techniques that would be used in the other occupied cottages, though, which included fortified underground bunkers. John

James was the under-sheriff in charge of the operation. Our job was to deal with the tunnels at the two main sites, Moneymore Cottages and Green Wood (where some of Dave Richards's ashes were due to be scattered).

As the tunnel experts, the members of my team were mainly from military and fire service backgrounds, so we were given far more information than others. Usually, a writ informs those being evicted of when the bailiffs will arrive, so they have time to prepare. We were told that the police had decided to take Green Wood covertly, however, without bailiffs. We moved in with them and they seized the site quickly – once more the day after dole money had been paid out and before anyone realized what was happening – and we quickly cleared the fortified bunkers.

Moneymore Cottages, the second target, was much more complex.

We had aerial photos of the site taken by the police helicopter, but all the windows of the cottage had been boarded up so there was no way of knowing what surprises lay inside. The protesters had been warned, though, that any injuries caused by booby traps would lead to a swift and robust response.

The operation started early in the morning, and we moved in to scope out the situation. The protesters had turned the building into a huge, creaking puzzle that needed to be unlocked piece by piece. It was impressive, but shockingly dangerous. There were five individuals in different parts of the 'puzzle' and each of them would have died if the building collapsed.

On the top floor a man and woman were locked into a barrel of concrete mixed with broken glass and rope. They were passive and allowed us to work around them. We put safety goggles on them when we started cutting, chatted and gave them hot drinks.

We asked them about the other obstacles in the cottage. When we had finally released them and removed what was left of the barrel and its contents, we removed the sheet it was resting on to reveal a man in the oil drum underneath with a bicycle lock around his neck. His feet were chained to a tyre full of concrete that was resting on the wooden roof of a chamber below. If the wood gave way, the tyre would fall and he would have been hanged. The whole set-up had to be supported with joists while we carefully cut his neck out of the lock. Again, as we carried out the intricate work the guy chatted amiably to us.

In the wooden chamber underneath there was a girl, Jess. She wasn't so pleased to see us. She was aggressive and spat and swore. We worked on her through the night during which I cut my hand open on a shard of glass trying to get into the tube she was locked into. The injury had to be reported and the police advised us that the site was then potentially a crime scene. I went off to make a statement in case the police wanted to charge someone. Although loads of protesters at all the sites we'd worked on had been arrested and charged, when they got to the magistrates' courts to answer public order charges, they were generally just getting small fines. This was another element to the protest, it was snarling up the justice system.

After I'd been bandaged up, I went back to work on Jess. Hours later she was removed, which then allowed us to get to the tunnel system in which we found another barrel with an urn concreted into the top. I lifted the lid. There were ashes inside and a couple of cigarette butts. We didn't know for sure, but it was likely that they were Dave Richards's ashes. We carefully cut the urn out of the barrel and handed it upstairs to the police, treating it with respect.

Eventually we got through to another hatch. A man was attached to it by a chain noose. He was one of the protesters from the meeting, a guy called Scotty. We'd heard through gossip on the site that when the operation began, he had taken a cannabis supply from the main camp with him when he went down the tunnel. We set up communications so Scotty could talk to the surface, and I went up to update ACC Green, who got on the radio and told Scotty he would be arrested for possession. I frowned because I thought worrying about possession of cannabis should be the least of his worries, but when he came off the comms he smiled and explained that Scotty would likely smoke it all now, which would make our job much easier. It was very common to encounter drugs on all the sites, but we used it to our advantage, as a stoned protester was generally easier to deal with.

Scotty was in a very tight space, and we used the same technique to remove the chain he was attached to as we had in Manchester, fixing it first so it didn't fall on him. When we got him out Moneymore Cottage had one final surprise for us. We found a pair of boots hanging out of a hole in the wall. It was Disco Dave, who'd been concreted in. We cut him out of his chamber while we chatted, then we had some good banter and a coffee afterwards. In all we were there for fourteen days.

The jobs kept coming. In September 1998, for example, we removed protesters who had taken over an old railway station in Oxford that was due to be demolished to make way for a business school. By then, SGI had fifteen people on the payroll and our reputation as confined-space and tunnel experts was growing internationally.

On that job the protesters had dug out a large void under the floor of the station in which six people were all locked into various barrels and concrete-filled holes. When we climbed in, the people lay there and stared at us silently. One woman had covered her face completely in red lipstick. It was eerie. As I went to work on one of the protesters, he told me to stay away because he had hepatitis C. At one point he unzipped his trousers with his free hand and urinated everywhere.

As we were working underground the bailiffs were supposed to be clearing protesters from the roof and ceiling. When we'd finished our area and were getting ready to vacate the building ready for demolition, we heard a voice.

'Oi! You've left me up here.'

We checked and found a protester locked in behind a chimney stack, which showed how every nook and cranny needed to be checked thoroughly before the bulldozers were called in.

Between 1996 and 1998 we worked constantly on environmental protests and developed a good relationship with the protesters. We were frequently thanked by them for our safety procedures and professionalism, for always ensuring their welfare and supplying them with hot drinks and medical care when it was needed. The rapport we built with them allowed us to understand their methods and strategies – and also to find out where they were heading next. This mutual respect and open communication directly contributed to our 100 per cent safety record.

As the protest movement grew and became better organized, so did our operational capacity. I invested in the latest and best equipment and training. I set up a rope rescue team and bought boats so we could run a marine unit, which was deployed to protect movements of nuclear material to and from Sellafield. I always had a vision to establish an underwater operation, using divers for search and rescue. I kept a close eye on any new technology that could be used to help us expand as a business.

In 1998 I went to the USA to research new underwater search equipment, which was years ahead of the UK with its search and rescue technology. On my return, I invested £80,000 on a new submersible remote operated vehicle (ROV), which is like an underwater drone that could be used to search the hulls of the ships and the bottom of quarries and lakes.

I had become frustrated with the lack of equipment available to search and rescue operators in the UK – no police forces used ROVs. I realized that with the right equipment we could not only improve the maritime services we were offering clients in the nuclear industry, we could also help the emergency services in cases where people had gone missing in bodies of water. It was frustrating for families and for search teams when people couldn't be found, and with this technology we could help with this. The ROV, I realized, would be good in deep quarries.

Through the contacts I had made in the police and during the work with the nuclear industry I was introduced to Mark Harrison MBE, then national police search adviser (PolSA). A PolSA plans and coordinates crime and missing person searches. Mark was the national lead and, after seeing the specialist skills we had, agreed to put SGI on the database for the National Crime and Operations Faculty (now the National Crime Agency), once we had been carefully vetted. We became the only team approved for police diving operations in the UK. Mark agreed to call SGI if he had any searches where our dive team or apparatus could be used. Around that time I was in contact with Cumbria Constabulary, where the chief inspector told me about an outstanding missing person case in deep water in the Lake District. I offered our services free of charge in order to see how the drone performed.

The missing person was a police officer who had disappeared while on a recreational scuba dive in Lake Ullswater several months previously. Cumbria Police knew the rough location where he had vanished but had not been able to find him as he was in forty-five metres of water. The local farmer who owned the land nearest the site allowed us to pitch our tents by the lake. We took two Zodiac boats onto the water, one to carry the large ROV and one to carry the generator we needed to power it. I had divers with me and the ROV required two operators: one to drive it and one to operate the cable. It was a boiling hot summer's day and it was hard to see the screen, so we went back at night when the temperature had dropped and I could clearly see the TV monitor.

We drove the boats out to the search area and dropped an anchor, set everything up, then dropped the ROV into the black water. As it descended the light from it in the water dimmed, then disappeared. All you could see on the surface was the glow from the viewing monitor. I watched the live feed as the ROV silently cruised to the bottom and landed in a cloud of silt.

We spent several days 'flying' the drone over the area. Looking for a body. It was monotonous and frustrating work and ultimately the search proved fruitless. This could have been a setback but I preferred to look at it as a learning opportunity.

The ROV, while great for targeted areas, was not so effective in murky water or in large areas where the location of the target was unknown. The search operation helped me realize the limits of the equipment and because of that I started to look at other technology that could be better used in wider search areas. The missing police diver, incidentally, was subsequently found a few months later by another recreational diver in a completely different part of the lake.

My search for useful equipment led me back to the USA and to the next major investment, which was a high-frequency side scan sonar (SSS), costing £35,000. The missile-like object was towed behind a boat and sent sound waves across the river, lake or seabed. The signals were then returned and processed on a computer in real time, building a detailed picture of the area below.

This piece of equipment gave us the potential to search for missing people, weapons, evidence and other lost items. It was an exciting, advanced piece of kit that wasn't being used by police dive teams in the UK at the time. The Royal Navy had it for mine-hunting but police searching for bodies or evidence in water relied instead on divers or volunteers using inaccurate equipment.

I had product training on the basics of how to use the apparatus, but in order to really understand its search and rescue capabilities I needed to get out into the water and clock up as much time with it as possible. So I conducted trials in reservoirs and canals with the SSS to learn about its full functionality and where it could be most effectively deployed. I needed to understand what a victim of drowning looked like versus a gun, shopping trolley or tree trunk. Would a decomposed body look different? What were the system's limitations?

We had a medical anatomical human skeleton that we used for first aid training who I named George. With the permission of Weir Wood Reservoir, in East Sussex, I took George out on one of our Zodiacs, dressed him in a boiler suit, tied a rope around his leg attached to a buoy, weighted him down and threw him over the side so he would sink to the bottom. Hoping no members of the public saw what was going on and called the police, I then started running search patterns past the marker buoy methodically up and down the reservoir, scanning the bed to see what image George created on the computer in front of me.

It took months of training in different bodies of water – I even took it to Spain and conducted trials in the Med with some of the team – but eventually I mastered using the SSS. I realized that the depth of vision it provided would allow me to locate a missing person in a lake or reservoir in hours, whereas without this equipment it would take weeks.

Mark came out on some of the training exercises I conducted with both the ROV and the SSS and looked at some of the scans from the test searches. He agreed it was a game changer, so the details of the equipment and its capabilities were added to the database.

We didn't have to wait long to use the SSS in an operational capacity.

Allison McGarrigle vanished in 1997 and was reported missing seven months later, in 1998. She was thirty-nine and lived in Rothesay on the Isle of Bute. In 1994, when she moved to the island after splitting from her husband, she befriended two men, William Lauchlan and Charles O'Neill. Unbeknown to her, they were paedophiles, and when Allison discovered that they were abusing a young boy, on 20 June 1997 she threatened to tell the police. She was never seen again. The pair murdered her, put her body in a wheelie bin and dumped it at sea. She was reported missing by her ex-husband. The men boasted to several people that Allison had been 'done away with' and was 'feeding the fishes' in the Firth of Clyde.

Mark called me in to look for the wheelie bin and her body after police had gathered enough evidence to indicate that she had been dumped at sea. We spent several days scanning a large area of seabed in Millport Bay in the Firth of Clyde, just off the island of Great Cumbrae. It was frustrating work because a wheelie bin is a big target and would have shown up clearly on a scan. I was disappointed when we couldn't find Allison but I was certain we had done a thorough search.

O'Neill and Lauchlan were not convicted of the murder until 2010, by which point they had also been convicted of a string of sexual offences against children. Although I'd not been able to find Allison's body parts, I could see every piece of fine detail on the seabed over a large area, and so proved that SSS worked as an effective underwater search tool that could be used for a range of purposes. Over the years it has proved to be an incredibly useful piece of equipment that has allowed us to locate many items of vital evidence as well as missing people.

Locating and recovering drowning or murder victims is distressing, harrowing and traumatic work, but to be able to offer families and friends of those who are missing some closure is an important and valuable endeavour. I realized this the first time we recovered the body of a man who had drowned while just enjoying the water. It happened on a lake in Ely, in Cambridgeshire. The victim, a thirty-two-year-old man, had been driving a speedboat pulling a water-skier. He wasn't wearing a life jacket and had an epileptic fit at the steering wheel. He fell out of the boat and disappeared under the water.

We were called soon after the accident, when it was evident he would not be found alive. His family had also arrived by the time we reached the lake. They were desperate for him to be recovered. It was a harrowing sight. Police officers were with them, and they were clearly distressed. I had seen death many times before with the London Air Ambulance and it is not something you ever get used to.

The friends who had been with him at the time pinpointed accurately where he'd gone in and as there was no current and the water was still I could be confident that the body would be near to the spot. We were careful to keep a low profile as we dressed in our diving gear and were shown the rough location of where to search. The friends and family knew what we were there for and I felt an acute sense of responsibility to conduct the job that we had to do with sensitivity and dignity. It was the first drowning victim I had been called to recover but there was no thrill or excitement. I felt a duty to the family and switched into professional mode. I tried to put emotions to one side, and to focus on the job in hand.

Having used the SSS in so many scenarios I knew it would not be accurate in that lake because the bottom was carpeted with weeds, which would distort the image. The lake was not deep and visibility was fairly good. We marked the spot where the man was last seen and one of my divers slipped into the silence of the water, slowly sinking to the bottom, where he concentrated on looking and feeling amongst the weeds, methodically sweeping a hand through them. It only took a few minutes to find him. He was lying face down, his hands still tightly gripping the reeds where he'd been trying to free himself. We gently untangled him and lifted him into our inflatable boat and took him to a quiet area of the lake where there were no onlookers. We carefully and respectfully placed him into the opened body bag. His eyes were open, staring at me, his hand still clenched into a fist.

It was a warm summer's day. We were in what should have been an idyllic location. The juxtaposition between the beauty and the horror was jarring. I remember being surprised at how quickly the flies found him. We zipped him into the bag and I went to speak to the family and friends who were, of course, devastated. I felt a profound sense of sadness for this man and his family and told them how sorry I was for their loss.

'We have found him and recovered him,' I said. They didn't need to know any of the details. No doubt they'd find out at a later date at the inquest that would need to be held. There were no more words that I could offer that would make their loss any easier. They thanked me and I walked away.

It was such a waste of life, and it struck me that his death was avoidable. If he had been wearing a life jacket, he would have survived. It was such a simple thing, such an inexpensive piece of safety equipment that could make such a huge difference to people who found themselves in open water and in trouble. Up to that point in my professional life I had been preoccupied with safety, whether it was going down tunnels or climbing buildings. I did everything I could to mitigate risk. It seemed logical to me that a life jacket was a perfect way to mitigate risk in water.

Some of his friends were nearby talking about having a barbecue to raise some money for his family. With the thought of water safety fresh in my mind I mentioned to them that they should consider donating some of the money to buy some life jackets. That was my first experience of drowning and I had many more in the subsequent years. At each one the pointlessness of the death hit me. I could never shake the thought that if more young people wore life jackets around open water, fewer people would die. Eventually that idea would turn into a campaign that I helped launch and roll out across the country.

Chapter 7

My life has always been about development and skilling up. If you stand still, you get overtaken, and the best way to ensure that you stay ahead of the game is to keep learning. I was a big advocate of training, and I made sure that the team were always properly skilled for all the jobs we did. Being skilled meant we were valuable, and the more skills we had, the more services we could offer.

Having established a dive team and purchased the ROV and the SSS, my former interest in diving was rekindled, particularly the idea of searching for things underwater. I'd always been fascinated with finding the hidden, whether that was old clay pipes in disused mines or eco-warriors in tunnel systems.

I had a natural ability to spot things and an inquisitive mind, which all stemmed from my father. When I was a boy we walked in the woods and he would always make a discovery. He was forever picking up coins and handing them to me, and he passed that awareness onto me. He bought me my first metal detector, a Beachcomber, and I took it on our camping holidays to Cornwall where I spent hours each evening on the beach after the holidaymakers had gone, scanning the sand for hidden treasure.

The first week I used it, I found £19 of dropped coins, which was a considerable amount of money for a child in those days.

When I put the SGI's underwater search team together, it was mainly made up of highly experienced ex-police divers, and I invested heavily in my own team to get them the appropriate qualifications. I also decided to take a commercial diver qualification and went up to Fort William in Scotland for a twelve-week course starting in late November, when the outside temperature was already freezing. Underwater search was one key element of the course, but I also wanted to learn to use tools and underwater cutting equipment. The course taught scuba and surface-demand diving, which is where a diver is connected to the surface by an airline, and the training was to take place in Loch Linnhe. It started in the classroom, but soon after we were walking along the icy pier over the water in preparation for our first dive. Our training began with an instructor called Jock, an ex-Royal Marine, who set us on our first exercise – mask clearing. This involved climbing down a ladder into twenty feet of water, lifting your mask in order to flood it, climbing back up the ladder without panicking to show the instructor your mask was full of water and then climbing back down the ladder, into the water to clear the mask using positive pressure. We repeated this training, progressing from simple scuba masks to full face masks.

As the course went on, we conducted searches under the pier, swimming under a grid of scaffolding practising search patterns with blacked-out masks to simulate searching in confined areas and in the total dark. We had to splice ropes on the seabed and bolt together flanges, just like in the film *Men of Honor*. There was a lot of study to do in the evening and written exams to pass, too. The dives got deeper, the course got harder, and the weather got colder.

Most of the other trainees were young men embarking on

careers in the offshore oil industry. It's fair to say they had youthful bravado on their side. I, on the other hand, had family responsibilities and a business to run. As the weeks went on, I got jittery. It didn't help that several weeks into the course someone drowned in a diving accident just off the pier we had been training on hours earlier. Although it had nothing to do with the dive school, it still felt like an omen.

The training exercises included simulations of the type of hazards that commercial divers faced. We practised on sunken Second World War tanks, which we climbed over and worked on using underwater cutting tools and other equipment.

On the first hard-hat dive we used a solid fibreglass helmet called a Kirby Morgan 'super light', which had a handle on top and was extremely heavy. Air was fed to it through a hose and there was also a built-in mic with a radio link to the instructor at the surface. The helmet was fitted onto your head by your buddies and clamped onto a neoprene neck seal. I was sent out from the bottom of the pier to clamber over the old battle tanks. I was enjoying my peaceful underwater surroundings when suddenly my helmet was filled with compressed air and was trying to shake itself off my head. I got on the comms and shouted: 'Get me up!' As I was being pulled to the surface, I tried to stop the air rushing in by turning the free-flow valve, but it was malfunctioning. When I got to the surface, my helmet was removed, and I told the instructor what had happened. He tried to play down the incident, which could have been catastrophic if the helmet had shaken loose. He said that there was nothing wrong, cranked the free-flow valve around to tighten it and told me to get back in the water. I refused and a heated exchange ensued. The maintenance man was called over from his workshop and later confirmed the valve had a bit of grit on the rubber seal which had damaged it.

The course carried on and we dived to greater depths with more complex tasks using a range of underwater cutting equipment on the seabed for up to three hours while wearing a suit fitted with hose pipes that leaked hot water to keep us warm. We practised rescuing other divers and I studied hard and worked at how to conduct underwater searches, swimming along a line on the riverbed, using your free hand to sweep the bottom. I learned how water moves and how objects in water behave and change, and I also learned that when you're underwater searching and there is zero visibility, it's best to close your eyes and let your sense of touch take over. I found I could concentrate in a dark, silent space, remaining totally focused. It took me back to the old days down the mines as a boy when I would sit in the dark listening to the muffled sounds around me. Sometimes in that world beneath the surface I found peace, swimming slowly and methodically along a rope, sweeping my arm back and forth, searching.

At the time, SGI was a growing company and we were already doing a range of different jobs. I took time out to do the diving course when there was a lull in the big protest jobs, but while I was away I stayed in constant contact with head office. I felt the training was important as I was adamant that SGI's capabilities would be built on cutting-edge equipment and advanced skills. I was constantly looking for the best bits of equipment and technology that would enable us to operate at the top of our profession and I valued the importance of regular training to prevent skill fade. We were expanding fast, as was the range of jobs we were asked to do. In any single month we could be called out by several private contractors or local councils to deal with protester work, and at the same time we could be asked by a police force to help in the search for a missing person in a lake or river. I had a team of skilled divers and a team skilled in protest removal and our reputation was one of professionalism and versatility.

During the Birmingham Northern Relief Road M6 Toll protests, another problem was brewing in South-East London on the site where Crystal Palace stood before it burned down. The Big Willow Eco Camp was set up to protest the building of a large leisure centre on parkland at the site. Eco-warriors had occupied the land for over a year and were living in fortified tree houses, tunnels and makeshift structures.

The operation to remove them in 1999 was called Operation Paxton. John Hargrove, the under-sheriff, was overall in charge of the operation working on behalf of the local authority, Bromley Council. Working closely with the Metropolitan Police and John, I carefully planned our part of the operation over many weeks.

I had a good idea of what we would be facing on the surface because I was given video footage of the site taken by the Met Police helicopter, but I had no knowledge of the extent of the tunnels.

The operation swung into action early one morning when bailiffs, police and my team were bused to the site in a fleet of hired Luton vans. When we eventually got to the site, breathless and hot, the media were there to cover the start of the eviction.

In the centre of the site there was a large tower full of protesters that the police and bailiffs would deal with. Our targets were an underground tunnel and two bunkers that had been topped by around 5,000 tyres set into twenty tonnes of concrete and built like nuclear fallout shelters. There were firefighters on standby because of the fire hazard the tyres presented. If the mountain of rubber had gone up in flames, the acrid black smoke would have created an aviation hazard and smoked out half of London.

We started in the bunker on the southern side of the site. The access shaft was under a bender (a makeshift shelter), which we removed. There was a large metal lid chained into the ground, so we dug down and cut the chain to release the lid, first ensuring there was no one underneath it. We removed that lid to expose yet another lid that appeared to be custom-made.

As we worked to safely gain access, the lid flew open and a frightened man appeared, gasping for breath and calling for help. We quickly pulled him out, put him on a stretcher and took him over to our medical tent that resembled a military field hospital.

He explained that he had been holed up deep inside the bunker and while changing the gas cartridge on his stove he didn't realize there was still gas left in the canister. As he unscrewed it, gas leaked out. He was using candles for light which he dived on to extinguish before they turned his living quarters into an inferno. Having averted death by fire he still found himself in an enclosed space that was rapidly filling with butane and made a bolt to get out. Unfortunately for him, the only escape route was the access shaft that he had deliberately made extremely difficult to get into, so between him and the surface there were five heavy steel trapdoors, each chained shut.

We were unaware of the drama unfolding in the tunnel as we could not establish any form of communications and the noise of the generators and air compressors running in the background muffled the sounds of the panicked man furiously unbolting the chains and doors below.

The protester would have been in there for several days before we could have dug down far enough to get him out. Thankfully, he'd saved us the job.

With access now granted, we ventilated the bunker and climbed down to ensure no one else was inside. When we were satisfied, we ripped the top off with a clawed excavator to see how it was made. This helps us build a picture for future operations.

The protester had a stove and an impressive array of survival equipment and supplies. The stove was made from a large Calor gas cylinder with a flat steel top for cooking and boiling a kettle, a hole had been cut in the front with a small, hinged door to put wood in. There was a chimney for smoke and a ventilation pipe going to the surface that was attached to a large plastic milk can with a twelve-volt computer fan tapped into the bottom. The fan would draw fresh air into his chamber.

I couldn't help but admire the detailed planning that went into his bunker but next I encountered one of the most spectacular obstacles I'd seen up to that point in my protester-removal career.

The protesters had first dug a huge hole in the ground, fifteen feet deep and around seven or eight feet wide. Into the hole they had lowered the front half of a car, which had presumably been purchased from a scrap yard and sliced in two. Once the car was settled at the bottom of the pit, they had built a three-tiered wooden framework around it, then filled all the soil back in around the outside of the frame, levelling the earth at the surface. What you were left with was an access hole on the surface and a wood-lined shaft leading down to half a buried car. Their ability to improvise as well as their sheer determination to create something more audacious than before was extraordinary. The manpower needed to create something like that was mind-boggling and would have required a small army of motivated people.

There were three protesters in the wooden shaft, each on a different level and all attached to the one above by chains, running neck to leg through gaps in the tiers. It was a vertical chain of three people. At the bottom, under the dashboard of the car, the protesters had created two other lock-ons.

The car was particularly difficult to navigate because of the narrow, confined and claustrophobic space we had to operate in. In those tight and dangerous spaces it paid to work with the protesters, rather than against them. After several days entombed underground, with limited capability of feeding or hydrating themselves and lying in their own waste, none of them particularly enjoyed the situations they were in, so the aim was to get everyone out safely and as quickly as possible.

The car obstacle had to be dealt with in a systematic way. From the top we dug down the narrow access point, opening the shaft out wider to get to the top guy, who was standing on the dividing platform between the tiers and whose feet were attached to a chain that ran through a gap in the floor of his tier. He was attached to the neck of the person below. Once we'd managed to get the chain off his feet and get him out, the process was repeated with the second man on the next level down and then the third, all the while opening the tunnel and shoring it. Eventually we got right down into the hole where the car was and cleared that.

Meanwhile, I was being pulled from pillar to post working underground and reporting to the police and under-sheriff on the surface, who were anxious to keep tabs on events below their feet.

Me as a boy, with Dad, by the wall plaque after a breakthrough into a new section of mine.

Walking back to the car following a Sunday in the mines, with a carbide caving lamp on my helmet.

Aged nine at the top of the forty-five-foot mine shaft. The area behind us is now London's M25 motorway.

My dad, *centre*, meeting with Dennis Musto and Robin Walls for the first time. The start of our underground odyssey.

A typical Wednesday night tunnelling through
the firestone to discover new sections of mine.

Me in the centre with no shirt, cooling off after a
ten-mile battle march in full kit with the Paras.

Aged seventeen on a family holiday with Mum and Dad in Greece.

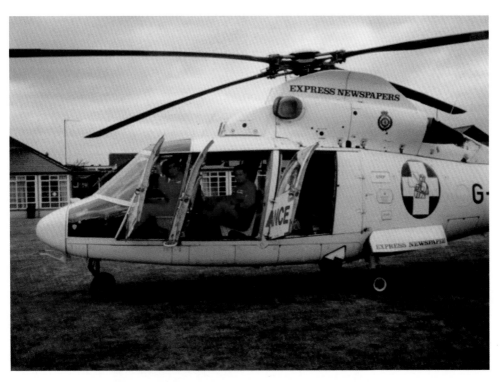

Riding in a London Air Ambulance helicopter in the 1990s.

With activist Swampy down the tunnel during the 1997 Honiton bypass protests.

Me and Swampy at his home in 2022, discussing my book.

The Birmingham Northern Relief Road protests.

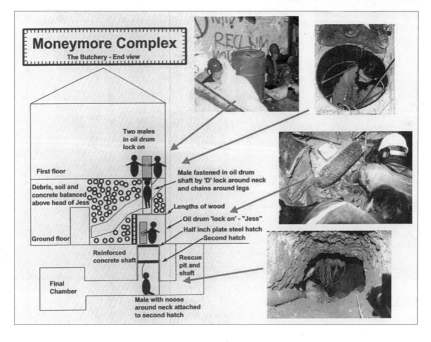

Moneymore Complex
The Butchery - End view

Two males in oil drum lock on

First floor

Debris, soil and concrete balanced above head of Jess

Ground floor

Reinforced concrete shaft

Final Chamber

Male fastened in oil drum shaft by 'D' lock around neck and chains around legs

Lengths of wood

Oil drum 'lock on' - "Jess"

Half inch plate steel hatch
Second hatch

Rescue pit and shaft

Male with noose around neck attached to second hatch

Digging out a female protester locked in at the bottom of
a sixteen-foot shaft at the Manchester Airport protests in 1997.

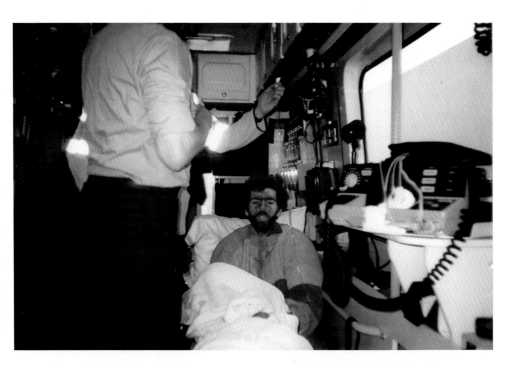

Disco Dave in the ambulance, having collapsed in shock after
I removed him from a tunnel that caved in ten minutes later.

As a guest of the United States Secret Service at the White House in 1999.

Diving training coming in very useful as I search the hull of a ship.

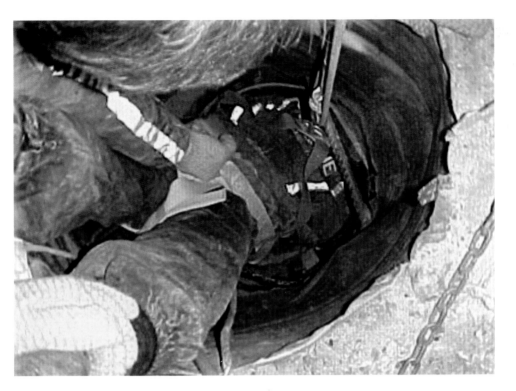

Being lowered down a shaft headfirst to release a chain around a protester's neck during demonstrations against the A130 extension in Essex.

Looking down the 750-foot flooded Llanharry mine, where I discovered the vital shotgun cartridge evidence in the Harry and Megan Tooze case.

Our ground radar equipment is some of the best in the world. Here the marks show a potential target in the Banaz Mahmod search.

My team, Specialist Group International, in training.

The last bunker with all the concrete tyres on top of it was an even bigger problem because we couldn't dig down through the concrete without mechanical aid, but if we went down that route there was the risk of the subsequent vibration causing a tunnel collapse. The only way to get in was to dig a long vertical shaft at the side, making sure it was deeper than the bunker, then dig a horizontal tunnel leading from the shaft under the bunker floor, finally digging up into the chamber from below. It was an audacious plan. The team decided to race each other from opposite sides to see who could get in first. When we finally got inside the bunker there was a pushbike on a stand in the middle attached to a car dynamo that was attached to two car batteries. The occupants could peddle the bike to recharge the batteries for their lights and ventilation fan. The bunker was extremely well made and lined with steel sheets.

The whole eviction took us nine days to complete without accident or injury.

It was at that job that the authorities started to gain a better understanding about the technicalities of what we did, and the skill set we delivered. We weren't just removing protesters; we were rescuing them. In turn, the protesters were evolving, too, refining their methods and finding more dramatic and effective ways to disrupt. The cheap, easy to buy bicycle D-locks that they favoured became straightforward for us to remove, so they moved on to Kryptonite locks, which were more expensive but much harder to cut. When we began encountering them, we had to buy better cutters to deal with them. It was a race to outwit each other, a game of cat and mouse played out below ground. My engineering training proved invaluable because I knew about tools and the properties of different metals. I needed to ensure my tools and equipment were continuously maintained and of the highest standard.

The protesters became divisive figures through the decade. They were derided in some parts of the media, hailed as heroes in others. They often had support from members of the communities that would be affected by the developments they were protesting. People generally did not want their local woodland chopped down to build roads or leisure centres. If I am honest, I sometimes agreed with their cause, for example in Crystal Palace, where the plan was to build a multi-screen cinema complex in the middle of a beautiful park in London.

Conversely, I also understood the need for some of the infrastructure projects. The Newbury bypass, for example, was being built to alleviate a terrible traffic problem. It is accepted today that the new road is a good thing and that it saved the villages it directed traffic away from. Nevertheless, my role in these jobs was to remain professionally detached from the controversy, which I always was.

I gave regular talks to the Police National Search Centre at Chattenden, in Kent. One day after a presentation, I was introduced to two American gentlemen by the chief instructor. Bill and Marcus were from the US Secret Service and invited me to Washington, D.C., to share my knowledge.

I was honoured to be asked and went out in February 1999. I stayed a couple of blocks from the White House and went to a military base where I presented on my protester removal experiences and detailed the extent of the disruption they caused to major infrastructure projects. I had lunch with senior officials and answered their questions – in the US, environmental protesters are classed as domestic extremists. Bill and Marcus then took me on a tour of the underground facility where all the Secret Service and presidential cars were kept. Later that afternoon we went into the White House and Bill gave me a tour, introducing me to other Secret Service agents. In the Oval Office, he told me that he used to look after President Reagan, who kept a stash of jelly beans on his desk. At night when he was doing security checks Bill would take a couple. One day Ronnie cornered Bill, put his arm around him, smiled and said: 'Bill, you've been helping yourself to my jelly beans, haven't you?' Bill said he was a lovely man.

In the UK my details were placed on the National Crime and Operations Faculty register, which was a list of approved and certified experts that all police forces in the UK had access to. It included highly specialized forensic scientists, and I was on the register as a specialist search expert.

My research into new search and rescue technology led me to look at ground-penetrating radar (GPR). It was first used on a crime scene in the UK at Cromwell Street, in Gloucester, in 1994, to search for Fred West's victims. The equipment used high-frequency radio waves reflected off underground items to map the substrata. It was mounted on a four-wheeled chassis and pushed over the area being surveyed like a trolley. The built-in screen showed the operator a scan of the ground underneath and showed any disturbances, anomalies or buried objects.

I realized the wide-reaching benefits GPR could add to my armoury of cutting-edge equipment; it could be used to locate protesters' tunnels, buried weapons, murder victims, stolen items and hidden chambers. In 1999 I purchased one for £33,000. My team and I began testing and training the system, burying various items and becoming familiar with what they looked like on the scanner. I became adept at understanding how different objects and disturbances presented when the radar passed over them. I could identify where earth had been disturbed, where small items had been buried and where holes had been dug and filled in. I could work out what was a root and what was possibly something more sinister, like a limb, or a shotgun. Given the difference this type of equipment had made in the west, I hoped that it would provide an extra asset that the police forces who were using our services could utilize. I was fascinated with police work and from a business perspective I could see that this type of equipment would be in demand, particularly in criminal cases. Little did I know just how useful it would be and how much demand for my help in crime investigations would increase.

One of the earliest cases on which I was asked to use the GPR (but didn't need it in the end) was the tragic case of a missing newborn baby in Surrey. A mother suffering from mental health problems had given birth and the baby had disappeared. It was not known if the newborn had been stillborn or died after birth. Police suspected the mother may have buried the child's body in her garden, so I was given the grim task of running the GPR over the area. I found nothing on that occasion and, if I'm honest, I was glad, but the story didn't have a happy ending because soon after the child's body was found in a skip.

Then, in early 2000, I was contacted by the Metropolitan Police to help with a murder inquiry. Sharon Malone, a thirty-year-old mother of two young boys from Potters Bar, in Hertfordshire, went missing on 29 November 1999. Her husband, Garry, made an emotional television appeal for her safe return but then strangely left the UK a few weeks later to live in Spain, taking the couple's children with him. He said the strain of the disappearance was too much for him to bear.

In mid-March 2000, Sharon's decomposed body was discovered at a local beauty spot, wearing just a T-shirt, underwear, a watch and a bangle. She was found in a dried-up riverbed caught up in a metal sluice gate, which suggested that she had travelled downriver from a site further upstream, at a time when the river was full. A post-mortem found she had fractures to the front and back of her skull and, in particular, a depressed fracture to the back of the head.

Because of the lack of injuries to her limbs, the coroner later ruled it was unlikely the head injuries had come from the passage of the body down the river and that the skull fractures must have been caused by some sort of weapon. There were no obvious defensive injuries to the hands, and the attack appeared to have been from the back, indicating that she'd been hit suddenly and violently from behind with a heavy blunt object.

At the time she went missing there had been heavy rainfall in the area and investigators surmised the body may have been buried in a shallow grave but then washed out by floodwater. I was called in to search for the likely disposition site.

The river she was found in, the Colne, had many tributaries, and police had a good idea of the area in which she may have been originally placed thanks to analysis of mobile phone data taken from her husband's phone, who was a prime suspect.

I began searching in the area and didn't need the GPR. Years of experience looking for hidden caves and tunnels meant that I had a keenly developed eye for out-of-place topography, and as I looked over the ground, I spotted a dip. It looked out of place and there was nothing natural in the area that could have caused it. Although it was dry, I could also see that it was an area over which water would flow during flooding. I got on my hands and knees for a closer look and bent down to brush some of the loose topsoil away. Underneath I could clearly see small roots that had been cut around the sides of the depression. This suggested that digging had taken place there at some point. The forensic archaeologist who was with us also checked the anomaly and confirmed that the ground had been dug with some form of tool or spade. I also ran the metal detector over it to check for any items of evidence.

With the likely shallow grave identified, I went back the following day to search the surrounding undergrowth. The vegetation was carefully cut back and in the middle of an area of brambles I discovered a ripped tarpaulin sheet with pieces of gaffer tape attached to it. This was taken away to be tested for DNA, as it was very possible this is what the body had been wrapped in before it was buried.

Like many of the cases I was involved in, there was no resolution for several years. In 2005, Garry Malone, who failed to return to the UK for his wife's memorial, was convicted of her murder and given a life sentence. It was discovered that at the time he killed Sharon their marriage was on the rocks, and he was having an affair. He was heavily in debt and facing the loss of his sons and financial ruin in a divorce he couldn't afford. Malone was trapped by mobile phone site evidence that showed that on the night of the murder he used his phone within hundreds of yards of where Sharon's body was found to make calls.

In another case at around the same time, I was approached by Staffordshire Police who wanted help with an unusual and gruesome puzzle that had appeared on their patch. They wanted to find out how long a human could survive inside a closed barrel. Council workers had discovered a sealed plastic barrel beside a country lane in Pendeford, near Wolverhampton, in 1999. When they looked inside, they were horrified to discover the badly decomposed remains of an adult, wrapped in a duvet. A pathologist later estimated the remains had been in there for three weeks. The body belonged to Floyd Dodson, a drug dealer from Amsterdam who was involved with drug mules smuggling cocaine between South America and the UK. He was so badly decomposed he could only be identified by credit cards and a notebook found in one of his pockets. It later transpired that he was lured to a meeting with gang members near his home, then tied up and beaten. The attack was thought to have exacerbated a heart condition caused by his drug use. It was not known whether he was alive or dead when he was wrapped in the duvet and stuffed headfirst into the industrial barrel, which was loaded onto a truck and driven to Staffordshire.

As a confined-space expert, I was contacted by the senior investigating officer (SIO) in charge of the case who wanted to determine how long Floyd would have survived on the journey if he was alive when placed into the barrel.

In hindsight me and my team were being called upon to provide a range of different services and skills, but at the time I took it all in my stride. I developed SGI's capabilities strategically to provide a versatile set of skills that would be useful for all the emergency services, and as the range of things we could do widened, so did my problem-solving skills. If a police force had an unusual request that involved an element of search, rescue or confined-space expertise, I could usually come up with a solution.

I enlisted the help of a couple of my team at SGI. Scotty and John were both former SAS paramedics, so with them we set about conducting some scientific human field tests. We used the same type of barrel, the same size, dimensions, design, make and material and took it in turns to scrunch up inside the drum with an oxygen metre, a torch and a spreadsheet. We were each sealed in, with others on standby outside with oxygen and a medical kit. The aim then was to breathe down the oxygen inside the barrel, measuring and marking the levels every fifteen seconds until we could ascertain how long it would take to suffocate, without suffocating ourselves.

Floyd had been quite a big guy and would have taken up more space in the barrel, especially wrapped in a duvet. My oxygen depletion tests did give a good indication that had he been alive when he set off on his final journey, he wouldn't have got past the South Circular before the air ran out. It was bone-chilling to think what a horrible demise he would have endured.

I provided a full report on my findings, and this was used by the police to help them build their case and understand what was likely to have happened to the victim.

Again, it took years for justice to be done. In 2003, Nigel Brade, who was described in court as an 'active and central member of an international drugs gang' was sentenced to only six years for Floyd's manslaughter.

Chapter 8

By the turn of the Millennium I hadn't seen Mandy and the girls nearly enough over the past few years and some family time was well-needed. It was time for a treat, so we flew on Concorde to Barbados for two weeks of relaxation and fun in the sun.

On my return home life got back to normal – when I say 'normal', I mean my type of normal. I was suspended by my feet upside down a narrow shaft sixteen feet underneath a Second World War pillbox in Essex. Below me, under a metal cover, there was an eco-warrior with a chain secured around his neck, which I was attempting to remove.

The protest camp was in Gorse Wood, near Rettendon, and had been set up in 1999 to delay the building of a new section of the A130, which, at the time, ran through a village and had become an accident blackspot. The proposed new route would take traffic away from the village but to construct it would require the destruction of woodland. The protesters were in situ not only to save the trees, but they also suspected the road was part of a bigger plan to create an extension of the M25 through rural Essex and Kent, which would then lead to more building and encroachment on the countryside.

Unlike at some of the other camps we'd worked on, the protesters at Rettendon weren't thought to have a large amount of local support. Indeed, many locals welcomed the new road and most of the anti-road activity was from groups in other towns. The camp itself was mainly occupied by protesters from elsewhere, and I knew many of them from previous protests. One in particular, 'Disco Dave' Dragonetti, had pedigree when it came to tunnelling. By the time the eviction notice was issued in November 1999, the environmentalists had been fortifying their tree houses and digging in underground for many months. And by the time my team and I arrived on site, they had created a fiendishly tricky tunnel. It was one of Disco Dave's best and it took us six weeks to clear it and close it down.

The intelligence we received from the authorities led us to an old Second World War pillbox, and it was suspected the tunnel entrance was hidden within it. Before we could get into it, though, we needed to dismantle a tall tower made of wood that had been built over the top of it, to which one protester was attached. Once that had been dealt with, we secured the area and set up our camp with our equipment, our medical tent and our all-important catering facilities. I had come to realize over the years that our tea urn was one of our most important pieces of kit, because a hot cup of tea always won over the protesters. The offer of a tea or coffee helped build trust, and when the protesters trusted us and realized we were not going to rough them up, they were easier to deal with. Tea was the lubricant that oiled our work.

The one entrance to the pillbox had been sealed with two reinforced steel doors and the firing slits had been filled in with concrete. We didn't know what was inside and while there was no suggestion of booby traps, we still wanted to see what we were up against in the interests of safety. Experience told me we couldn't just drill through the doors or slice them with angle grinders, in case someone was attached to them on the other side. So we carefully chiselled the concrete out of one of the slits until it was big enough to pass a digital flash camera through to take some shots. The images revealed there was no one inside and that the interior was full of soil.

'Where the hell are they?' I mused as I flicked through the pictures.

We spent several hours taking off the two reinforced doors, then I went in first and looked around. The whole floor was about three feet deep in soil. On top of the soil there was a car tyre with the flywheel of a car engine concreted into the top of it. I carefully lifted it up on one side and looked underneath. There was a long chain fixed to the underside that was dangling down a deep, dark shaft. From previous experience I thought that the chain might be secured around a person's neck. I called down. There was no reply. I gently pulled the chain, which was slack, then continued to pull it up until all of the chain was on the surface.

I rolled the heavy cover aside to reveal a narrow shaft descending into the darkness that had been fashioned by sinking oil drums with their tops and bottoms removed into the earth. Concrete had been poured down the sides to make a secure shaft. This would have taken months of hard graft; digging through the floor of a World War pillbox is no easy feat.

Suddenly, from somewhere down below I heard a voice.

'Don't lift the lid. I'm attached to it.'

From the top of the tube, I shone my headlight down to reveal a metal cover about twelve feet further down. The voice came from under it. It dawned on me how difficult it was going to be to get to the person under the cover because the oil drum tube was extremely narrow.

We formulated a plan. First my team cleared all the soil inside the pillbox, so we had a flat, clear surface to work from, then we used GPR around the outside to get an idea of in which directions the tunnels were leading and how long they were. In doing this we discovered that there was a much larger tunnel leading away from the access shaft.

We set up a rescue tripod over the hole. This was a three-legged apparatus commonly used to enable rescuers to lower ropes and lines into holes. It is placed over a hole or shaft and has a winch attached to its apex, through which rope can be lowered or raised, thereby allowing the rescuer to safely descend into shafts, manholes and other confined spaces.

Next came the hard part. The shaft was only just wide enough for one person, with no space to bend down to work in. The only way I could access the cover and get to the person underneath was to be lowered in head first. My team secured a webbing strap around the outside of my boots, tied my feet together with it and attached a carabiner onto which a line was secured. I was then carefully lowered into the steel tube.

The protesters had created ledges in the side of the shaft at intervals, which were hand and foot holds, so these gave me something to hold on to and lever against as I was slowly lowered down and began the difficult job of trying to get access to the underneath of the cover to be able to release it.

As I was upside down, all the blood rushed to my head, so I could only work in short bursts. I had to be winched back to the surface repeatedly to recover, which made me realize that mine was probably the only job in the world that required a person to be suspended upside down and lowered into holes.

Eventually I managed to tip the lid slightly, then I reached underneath and felt the chain and the protester it was attached to. He was in a tiny space, and feeling around I touched his head. It suddenly occurred to me that if he grabbed me and handcuffed me to something, I was stuffed. That made me very nervous.

'I'm just feeling around to work out where the locks are and what we're dealing with here. Do not grab my hand,' I said.

I encouraged him to talk, to divert his attention from any thoughts he may have had of locking me on.

I felt along the chain, and I managed to work out that he was shackled by the neck, he wasn't padlocked but had a small carabiner holding the chain together. I slowly managed to unclip the carabiner to release the chain, at which point he immediately said: 'See you later, Pete,' and scuttled off into the innards of the tunnel system. I knew that would happen. It would have been hopelessly optimistic to think he'd follow me up and leave. Instead, he scarpered into a deeper part of the system where he would make himself another obstacle to be overcome later.

After another 'blood readjustment break', I went back down and removed the cover that the protester had been attached to, revealing the main tunnel.

'We've got a big job here,' I called up.

At the surface, Big Pete and I discussed the situation to work out the best plan of action. A larger-than-life character and a practical joker, he had been having a good laugh about my inverted predicament, and while he was careful to make sure everything was safe at the surface, I think there was probably part of him that enjoyed the sight of me being strung upside down and lowered into a hole. We had to laugh in situations like that, it helped the team bond.

The oil drum tube was the only way down, but it was too narrow to get a stretcher through in an emergency. We knew from the ground radar images where the start and end of the tunnel was, and we couldn't dig into their tunnels in case they collapsed, so we decided to dig a rescue shaft about six feet away and then dig a horizontal tunnel to intercept theirs.

Everyone was briefed and over several days we excavated our own safe, shored, wide-access shaft that ran parallel to their narrow oil-drum death tube. That gave us access to their tunnel and allowed us to start digging them out.

Their tunnel was sectioned off with eleven steel doors that had been custom-made by a local blacksmith who supported the protest. We had to dig out each door, take it to the surface, and shore each section as we progressed. We worked in shifts; it was like a factory line, with each of us assigned to a specific task. Each morning we discovered the protesters had dug deeper throughout the night and backfilled the tunnel with piles of spoil that we had to clear. They also left us bags of their excrement and urine to get rid of.

It became a race underground to catch up with them, but with our superior equipment, manpower and skills, day by day we got closer.

The job finally ended in March 2000, when we removed the last five protesters from the tunnel after thirty-four days underground.

Although Disco Dave was instrumental in creating and building the tunnel, he wasn't in it and instead remained on the surface and acted as their liaison officer. He spoke to the media, giving them insights into what the protesters would be going through and how they would be feeling. He told the reporters they were depressed at having been removed, but glad the dig-in was over.

'Their legs will be stiff because of the lack of circulation, and they will be very disorientated,' he explained.

Their aim had been to cost the developers as much money as possible and to persuade them to rethink the route and save the surrounding countryside. They achieved their first goal; the cost of clearing the camp was estimated at £6 million. Later that year, the construction of the road went ahead.

The A130 protest marked the end of a period of back-to-back road protests. Big road-building infrastructure projects began to scale back as the political agenda changed towards encouraging less car use and more use of public transport. As such, the eco-warriors did play a part in raising awareness of environmental issues. Manmade climate change and the reliance on fossil fuels became policy issues.

It wasn't the end of protest culture by any stretch, though. Instead, the movement evolved and the causes changed, but the same old faces remained dedicated to the underlying environmental cause.

At the same time our work was starting to evolve, too. We had been involved in ongoing sensitive work in the nuclear industry, and we were well respected by law enforcement and Customs and Excise, as we trained their rummage teams. We also had an experienced and well-equipped marine unit whose divers had searched ship hulls underwater using the ROV, and who would be called regularly by Her Majesty's Customs and Excise, as it was then known.

I didn't have time to sit back and take stock of how things were developing professionally and there was no great business plan. SGI was providing services that were needed and I was happy to go wherever the work took me. I had no idea of the direction in which it was about to go.

Chapter 9

One of the great things about the way my life and SGI had evolved was the range of jobs we were called upon to do. I had contacts in so many agencies by now (some of which were highly confidential, so I cannot mention them) that if something needed to be searched, we were the first port of call.

In December 2001, less than three months since 9/11, and when the world was on high alert for follow-up terror attacks, I was in Crawley and, being a bloke, I was doing last-minute Christmas shopping when my pager started to beep.

The message read: 'Contact the anti-terrorist branch. Urgent', and a phone number followed.

I dialled the number on my mobile.

'Your ears must be burning, Peter,' my contact Alan, a commander in the Met Police, answered. 'I'm in a meeting. We have a major anti-terrorist operation ongoing off the South Coast. There's a cargo ship potentially carrying nuclear, biological or chemical weapons. Special Forces and the Royal Navy have seized it and we need the holds searched. Paul from HM Customs is here with me, and he says you are the man – he says hello by the way. It's a fast-moving operation and we need you and your team here urgently.'

'Absolutely,' I replied.

The ship was called the MV *Nisha*. There had been a tip-off that it might be carrying deadly material along amongst its cargo of sugar, as its route had taken it from Mauritius with a stop in Djibouti, close to the suspected Al-Qaeda havens of Somalia and Yemen.

The intercept and search operation included commandos from the Special Boat Service (SBS) and Special Air Service (SAS), along with bomb disposal experts and anti-terrorist police. The Royal Navy Type 23 Frigate, HMS *Sutherland*, had intercepted the ship off the Sussex coast at 5.30 in the morning of 21 December, launching Special Forces teams in RIBs (rigid inflatable boats) to board it. They were supported by Lynx Mark 8 helicopters, carrying snipers to provide cover along with two Chinooks, from which other Special Forces commandos fast-roped onto the deck. The ship was stormed so quickly that the crew would have been taken by complete surprise.

Anti-terrorist police and bomb disposal officers then boarded and performed an initial sweep but found nothing. The ship was moved and anchored a mile off the Isle of Wight in Sandown Bay, and it was at that point that we were called, as they needed someone with expertise in searching the holds of ships, which we had done several times for HM Revenue and Customs.

After I assembled a team, we whizzed down to Southampton docks where a makeshift control room had been set up with desks assigned to each of the parties involved in the operation. There was HM Coastguard, HM Customs and Excise, Hampshire Police, the Metropolitan Police, the Royal Navy, the Army and the security services. Parked outside were the big white unmarked trucks belonging to the SBS. I walked into the large shed for a briefing, after which we waited to be deployed onto the ship.

The area was a hive of activity. It was exciting to be at the heart of something so well organized and potentially so important.

However, at midday the entire operation suddenly stood down and was quietly wrapped up without explanation. We were thanked, then we drove back to our base. Even though our services were not required in that instance, I did reflect on the way home just what a thrill it had been. It was about as near to James Bond as you could get. The operation had also sent a message to any potential terrorists just how quickly and firmly the UK will deploy Special Forces and other agencies to neutralize any threat.

Indeed, my professional life was starting to take me to places I had never dreamed off. Increasingly, I was collaborating with police forces and had built up a network of high-level contacts. I'd never been a big fan of crime shows or harboured any ambitions to be a detective or an investigator, but as my services were picked up by different police forces, I became fascinated with the world of forensic and crime scene searches.

In the spring of 2002, I was called to Lincolnshire by Mark Harrison, the national PolSA, to help with the search for Laura Torn, an eighteen-year-old living in the pretty village of Owston Ferry. She enjoyed riding her pony and wanted to be a police officer but had gone missing after last being seen leaving the Crooked Billet pub in the village market square in the early hours of 27 April, after celebrating passing her driving test. She was involved in an on-off relationship with an older man, Guy Beckett, the landlord of another pub, The Red Lion. He was the main suspect in the case.

Initially I got the call to join the search for her because one of her shoes had been found by the side of the river that ran through the town and police suspected her body, or evidence, might have been thrown in the water. I was initially wary, I thought that maybe the shoe had been placed by the river as a red herring.

Nevertheless, we put our boats on the water and started scanning the riverbed using sonar. We found nothing the first day and the following morning I was in the incident unit that had been set up in the village briefing the SIO when Beckett walked past. Although he was the suspect, he hadn't been arrested at that point as the police didn't have enough evidence.

Beckett had been acting strangely before Laura disappeared. The CSI team on the case had identified scuff marks on the wing mirrors of his car that matched a narrow gangway down the side of the pub, suggesting that he'd tried to manoeuvre the vehicle to the rear of the property for some reason, possibly to load a body and take it somewhere to be dumped.

The SSS proved its worth in allowing us to confirm that Laura's body was not in the river and therefore focus could move to other areas. We were asked to continue the search, particularly in places where Beckett was known to have been in the time following Laura's disappearance. He was seen at a KFC branch a few miles away, so I drove around the area looking for evidence or possible disposition sites in lay-bys and remote tracks, as well as places a car could park. A high number of murder victims are deposited within fifty metres of a lay-by or a trackway, which is quite a long way when carrying or dragging a body. I pulled into one lay-by, searched a ditch that ran alongside and found a KFC box that could only have been a few days old. I was elated with my find and I rang it through to the SIO. My find, along with fresh information that came in, helped narrow down the search area and we were asked to focus on haystacks in the region, of which there were hundreds.

Eleven days after her disappearance Laura's badly decomposed body was found buried in a haystack in Misson, Nottinghamshire. Beckett was charged with her murder. At his subsequent trial, the court heard that he had become increasingly abusive when Laura rejected his advances and had strangled her after an argument in The Red Lion. He was jailed for life after admitting murder.

The level of trust I was given by the police, and the confidential information I was provided with by them may well have appeared unusual to outsiders unfamiliar with how investigations work, but crime scenes often involve a range of people, many of whom are civilians who have been carefully vetted, which I was. For example, scenes of crime officers are civilians employed by the police, and the police would call upon a range of experts when they need specialists in different areas, such as forensic archaeologists, botanists who can work out where a piece of evidence, soil or body part has been by testing the organic matter on it, medical experts who can map the patterns of blood splatters and psychologists who can analyse criminal behaviour.

Increasingly I was being called upon by the regional crime advisers from the National Crime and Operations Faculty at Bramshill. The regional advisers and the national police search adviser develop the search strategy on missing person cases, and I was being brought in on an increasingly frequent basis to locate evidence and human remains. I felt honoured to be able to help, and when I was called to a job I did everything I could to get a result. If I needed to stay all night, I would, and I would get frustrated when I didn't find what I was looking for. Because of this I was invited by Keir Hartley, one of the top trainers at the Forensic Science Service (an executive agency of the Home Office) to attend training in forensic awareness.

I had read a lot on the subject by that stage, and I had also learned quite a bit about forensics when I was on carefully controlled crime scenes, but the course widened my knowledge. It taught police officers, crime scene investigators and anyone involved in crime scenes the basics of forensics, aiming to give students an awareness of how to behave in a crime scene, how to handle evidence and how actions can contaminate evidence. It covered subjects such as DNA transference and fibre contamination. The classroom included a simulated crime scene complete with blood splatters and hidden evidence. I learned how a single strand of hair from the head of an investigator can lose the case for the Crown, how an item such as a pair of glasses put down on a table in a crime scene contaminates the evidence and how a murder scene is managed.

The course gave me a valuable awareness of how our work could impact a scene, in particular ensuring that all our tools were sterile. For example, if we were digging a potential disposition site in Surrey with shovels that had soil on them from a job in Manchester, the cross-contamination could throw an investigation in the wrong direction.

The importance of forensic awareness was highlighted a while later when we were called to assist in a major undercover drug-smuggling operation. We were part of a big criminal investigation into which a lot of police resources had been invested. For operational reasons I can't provide too much detail, needless to say the gang the police were after were dangerous. Police had been given a tip-off that the smugglers had buried a consignment of cocaine they had smuggled into the UK in an area near a school field. My job was to use GPR to locate the buried stash, dig it up and replace it with an identical surrogate that had been laced with smart grease, an invisible compound with a unique DNA marker that is transferred to anyone or anything that touches it.

I did the job with Chris. We took sterile tools that had been carefully cleaned and sealed in bags to make sure there was no risk of soil contamination, and we went undercover. We had to look as inconspicuous as possible, in case the criminals were watching from a hidden vantage point, so we disguised ourselves as struggling workmen. We used a beat-up old van to get to the job, dressed in work clothes and even wore flat caps. The attention to detail was such that I made sure I looked unkempt and I grew several days' stubble. GPR equipment is used on construction sites to survey ground for pipes, so when we talked on the job, as we were scanning the ground with the equipment, we discussed where the 'pipe' we were looking for might be, in case anyone overheard us. We also used different names.

We found an area that the radar showed had been disturbed, discreetly photographed it and dug down. A few inches down we discovered a large, carefully wrapped block of cocaine, which we photographed, removed and placed in an evidence bag. We calmly covered over the hole, taking pains to ensure the ground looked the same as it had before we dug. Then we casually carried on with the 'job' we were supposed to be doing before we made our way back to the van and drove a little way to meet our police contacts in a prearranged location, where we handed over the evidence.

The police later replaced the drugs we had removed with a surrogate and used our photographs to ensure that the disposition site looked exactly as we'd found it. Thanks to my forensic training I made sure the area was clear of any contamination from our tools, so as not to attract suspicion, or jeopardize any future prosecution. Undercover police then followed the shipment after it was collected and were able to tie individuals to the smuggling operation using the smart grease as evidence.

A few months after the course, I was called to assist in the search for a missing teenager in Wales. Jenna Brookfield Baldwin was fifteen when she disappeared from her home in Gwent in early September 2002. She lived with her mother, Desiree, who reported her disappearance to the police, and her stepfather, Mike Baldwin, who worked as a security guard at a chicken processing plant. In the weeks after the disappearance, Desiree received silent phone calls which were initially thought to be made by Jenna, and she became even more convinced her daughter was alive when she started to receive text messages, such as 'Mum, don't worry. I'm okay. I'm happy where I am'.

However, it transpired that it was Mike Baldwin who was making the calls and sending the messages in order to throw police off his scent; he had already killed Jenna and buried her in woodland several miles from their home.

Police suspected him from the start. There were inconsistencies in the statements he gave and a call he said that Jenna made to the house could not have been made as the line was dead. Nevertheless, investigators played along with his ruse to give the impression they were involved in a missing person's investigation, all the while building the case against him. We were called in by Gwent Police to conduct a GPR search of specific areas they thought likely we might find her body.

Before we started we were briefed by one of the senior officers in charge of the case. The briefing was concise and serious and we were told the basics of the case and our role within it. We were there primarily to find a body. As a father it would have been easy to allow emotion in – after all, who does not feel sadness, concern, dread, anger and the whole maelstrom of gut-wrenching feelings when they hear about a child's disappearance? But as a professional I needed to focus on the task and concentrate my energies on the search. At that point there was little that any of us involved in the search could have done to save Jenna, but we could help her family and try to get justice for her, and the best way to do that was to be as thorough as possible and conduct ourselves with laser focus, unencumbered by emotion.

Initially we were tasked with searching the garden of the family home, which we did with GPR. I could tell from looking that there was no disturbed ground but I wanted to be 100 per cent sure, so we did a meticulous radar search, because even if Jenna wasn't buried there, some evidence might have been, such as a murder weapon. After searching the garden we were tasked with looking further afield and were asked to identify and check any places where a body may have been dumped.

In woodland near the house I noticed a manhole cover and immediately remembered the story of Lesley Whittle and the Black Panther, aka Donald Neilson. Lesley was a seventeen-year-old heiress to the Whittles coach company fortune and Neilson kidnapped her from her bedroom in 1975 hoping to receive a ransom. He kept her hidden and tethered by the neck on a ledge in a drainage shaft, where she was eventually found dead, having either been pushed or fallen from the ledge. Drain systems are common disposition sites and this one was near enough to the house to merit a search. Bodies aren't generally carried far; they're usually found within fifty metres of tracks in woodland, and in crimes of passion, where there is no planning, the killer usually panics and tries to get rid of the body as quickly as possible.

Sometimes a search team will lift a cover, shine a torch into the shaft and then move on if they don't see anything. A smart killer will drop a body in, then go in after and drag it down out of sight. With all this in mind we went into the vertical access shaft and crawled along the horizonal drain to be certain.

The investigation was live and as police found new information they acted accordingly, reassigning assets where necessary. After a few days they received some new information and we were asked to a nearby quarry to search for Jenna's mobile phone, which they believed may have been dumped there. We took our ropes to the area and set up lines down the quarry face. Starting at one end we descended top to bottom, conducting an inch-by-inch finger search as we dropped. It was painstaking, meticulous work. When a climber reached the bottom, the rope was moved a few feet along and the process began again. I don't know how many descents we made but it took hours and hours. Remarkably, we did find a mobile phone and when the shout went up that a target had been acquired, I allowed myself a nod of satisfaction at a job well done. The phone was carefully placed in a forensics tube and sent off to be analysed, where, unfortunately, it was confirmed that it was not Jenna's.

We were there for a week and the whole community was

supporting the search. We had an operations trailer set up in a car park in the village and one day one of the locals bought us a freshly made shepherd's pie, which I thought was a lovely gesture.

One of the last areas we were tasked with searching was a lake up in the hills that was on Baldwin's route to work. We had taken an all-terrain vehicle and two quad bikes, which allowed us to get our equipment to remote locations quickly, and they proved invaluable in the tricky terrain.

Local folklore rumoured it was bottomless, but when we scanned it using the SSS it was found to be only six feet deep in the middle. In one hour, we were able to conclude that Jenna's body was not there, the same search using divers would have taken several days. On occasion lakes have even been drained to search for bodies, which is expensive and has huge environmental impacts.

We searched for a week and didn't find anything significant, but by November the case against Baldwin was getting stronger and stronger. He made lots of mistakes. Soon after Jenna disappeared, for example, he bought a new car and told police the old one had broken down. They tracked it down and found there was nothing wrong with it. He was arrested and admitted that he had killed Jenna. He said she accidentally fell down the stairs during an argument. Twelve weeks after Jenna's death he led police to a remote woodland mountainside area where he'd buried her in a shallow grave.

While a motive was never established, one line of inquiry suggested Baldwin made sexual advances to his stepdaughter, which she rebuffed. I learned later that one of the pieces of evidence produced at trial was that Baldwin had gone to a local hardware shop and bought a shovel using his credit card the day after she went missing. In 2003 he was found guilty of murder by a jury and handed a mandatory life sentence.

A few weeks after our search operation in Wales, I was called to take part in a completely different type of job by the Office of the Deputy Prime Minister (ODPM), who at the time was John Prescott.

The Fire Brigades Union (FBU) was threatening to call a national strike, wanting a 40 per cent pay rise. Mr Prescott, negotiating on behalf of the government, called the demand a 'fantasy'. As talks continued, the government organized a contingency plan in the event of a strike. The Army would respond to emergency service calls using old fire and rescue appliances called Green Goddesses. The ODPM wanted to know if SGI could provide support in the event of any road traffic accidents, or incidents involving confined spaces or rope rescue.

At this point SGI was involved in protester removals, water search operations, crime scene searches, specialist forensic tests and marine operations for the nuclear industry. It was quite a CV. I reflected that my boyhood dream, to recreate *Thunderbirds*, had come true. It was incredible.

The Fire Service cover, however, was a tricky decision morally because I fully supported the Fire Service and had worked with them on many occasions and didn't want to break their strike. However, I also understood that my team and I had skills and experience, not to mention equipment and resources, that could be invaluable in helping the public and could ultimately save lives in the event of a strike. And in the end, it was this that won out. I agreed to help, then talks between the government and the union broke down. On 14 November 2002, members of the

FBU staged the first firefighter strike for twenty-five years. It was a forty-eight-hour walk-out, during which time SGI was on round-the-clock alert. My team stayed at our Dorking base with our vehicles pre-loaded with all the equipment we might need. We used the time to enhance our skills, and former firefighter trainer Big Pete taught us road traffic extrication skills.

It didn't take long for us to get our first call.

There had been an accident on a country road and a girl, in her late teens, had been knocked off her bike on a sharp bend and was trapped under a car. We raced to the scene and found the girl directly under the car, face down with her mangled leg trapped against the hot exhaust, which had burned a large hole in her skin. An emergency doctor arrived at the same time, assessed her injuries quickly and quietly told us that he thought she might lose her leg.

We needed to get her out from under the car as quickly as possible and the best way we could do that was by lifting it off her and cutting through any pieces of it that we needed to get out of the way. Luckily, Big Pete was with me on the job.

We secured the car in place with a large strap, which we tied to a tree, then used wedges to stop it from moving in any direction. We then carefully jacked it off her and got her out from under the vehicle. Once she was free the doctor took over, got her into an ambulance and took her away. A Green Goddess turned up shortly after. I never heard what happened to her after that, but I know we saved her life and for that, agreeing to help against the wishes of the FBU was worth it.

Chapter 10

Some jobs stay with you, no matter how much time passes. Something happens, some horror, some display of inhumanity, and the event gets branded on your mind forever, like a stain.

I know I'll never forget Adam Morrell. I never met the young lad, but a terrible fate convened to make our paths cross in November 2002.

By the time my team and I were called by Mark Harrison on behalf of Leicestershire Police, Adam was dead. Pieces of him had been discovered all around Loughborough. First it was an arm. A woman walking along the Grand Union Canal noticed a suspicious-looking black plastic bag wrapped in silver tape floating in the water. It looked odd so she fished it out. It was a weird shape and when she peeled away the top, she saw fingers and a hand poking out. Next, both legs were found in a similar bag under a hedge. A torso was found at the end of someone's back garden. A bag of clothes was discovered dumped in an electricity substation.

Adam was a troubled fourteen-year-old. He had been bullied at school, which he dropped out of and was living with his father, a barman. He was trying to get back on track and was reported to have been doing well at the pupil referral unit he was attending before his death.

When he met an older teenager, policeman's son Matthew Welsh, who offered to teach him self-defence, he jumped at the chance. He began hanging around with Welsh, his girlfriend, Sarah Morris, and their older friend Nathan Barnett. Welsh and Barnett were trying to set up an employment agency from a house in Havelock Street, Loughborough, and had recently moved into the property. Morris also moved in after meeting Welsh in a nightclub. Adam would visit the trio after finishing at his referral centre in the afternoon and on a few occasions stayed there overnight. It was thought he befriended the gang because he was a bit of a misfit and wanted someone to hang out with. He was described as a lovable rogue who wanted to be part of a group of friends. On 14 November Adam called his dad to stay he was staying with friends that night, then met with the older trio who had been drinking, smoking cannabis and taking ecstasy.

Events took a turn for the worse when Adam threatened to tell police about their drug use. The gang set upon him and started to kick and punch him. What followed next was an un-imaginable two-day ordeal in which Adam was repeatedly beaten and tortured. A subsequent trial heard that his head was kicked 'like a football', and that his face was so swollen one of the attackers said he looked like an 'alien'. He was stamped on and boiling water laced with sugar was poured over him. (This is a particularly cruel and painful form of assault, known as 'prison napalm' – the sugar is supposed to make the compound stick to the skin, thereby increasing the burn.) Music was turned up to drown out his moans. Morris later told police that late in the afternoon of the second day of the attack she heard Adam being choked to death and that his body was then dumped in a bath where it was dismembered with hacksaws.

I knew none of this when I was asked to search the canal where some of the body parts had been found. All I knew was that a lad had died and that his body had been disposed of in the most horrific way. His head and his other arm had yet to be discovered. It was such an awful case and I wanted to find him, or the remaining parts of him, which sounds macabre, because who in their right mind would want to make such a terrible discovery? But as in the case of Jenna Baldwin, my role was to help the investigation, and by doing that I would also help the family and loved ones. My feelings about the suspects had to be put aside. Emotion was not an option, so I pushed down any revulsion I felt and focused on the job in hand. I was not there to make judgements.

With this in mind, after a sombre briefing, we set up by the side of the canal that we had been asked to search. It wasn't a huge operation; there were other police officers carrying out inquiries in the town but we were the only people tasked with searching the waterway. There wasn't much conversation between us. The difference between this job and the protest removal jobs was stark. There was no banter, no joking, just very serious, quiet and meticulous equipment checks.

We used the SSS to meticulously map the bottom of the canal for a few miles, taking sweeps up and down. I was focused on the screen, looking for anything that appeared out of place and unusual. Once I was given an area to search I was left to conduct the operation and direct my team. I was the trusted expert and police left me to do what I needed to do. Initially the search area was clear save for a few old traffic cones, a shopping trolley and two bikes. But it wasn't long before I identified several interesting items under a road bridge, where the water was only waist-deep. I stared intently at the screen; there were two or three roundish anomalies that showed up that I couldn't identify. My dive supervisor, Robin Jarmain, was a highly experienced ex-Sussex Police diver. As it was shallow, he reached into the water and slowly swept the canal floor with his hands. Then he suddenly stopped.

'I have something,' he said. We marked the area and I went to get the scenes of crime officer – who is there to look for fibres, fingerprints and other evidence and pack and document what they find, which is then sent to labs for scientific analysis. He was on the bank nearby and brought a sterile plastic sheet over to us and placed it on the ground. Robin then lifted the object from the canal bed and carefully passed it to me. It was wrapped in a black bin liner and was perfectly bound with silver gaffer tape. Everything told me that inside the package was the boy's head, and in that moment I felt my blood go cold. I shivered, not through any sense of fear, but through the awful realization of what that child must have gone through and the utter disrespect that had been shown to his remains. It was incumbent on me to at least give him some dignity and I very gently placed Adam's head into a sterile cardboard evidence box. What stayed with me was just how heavy it was, like a football full of water. I don't know what I was expecting, but I didn't realize a small teenager's head would weigh so much. Shocked and emotional, with tears running down my face, I walked away, staring into nothing. The only saving grace was that the whole gruesome scene was shielded from view because we were under the bridge. It was a Sunday, and above us life was going on as normal; throughout the town – apart from the pockets of activity at the house where the attack happened, which was sealed off – you wouldn't have known something so awful had taken place.

The head was sent off to the morgue where a pathologist would examine it. Their job was to analyse cuts and marks and match them with any potential weapons that had been recovered. Adam's body would then be put together like a jigsaw puzzle to build a picture of what had happened.

Once we'd finished the job, we packed up and drove back to Surrey, a long journey that was largely quiet and sombre. Over the following days I sometimes woke at night and could still feel the weight of the head in my arms. It was unusual for me to get emotionally involved in the cases I was working on, and up to that stage I had been able to compartmentalize my feelings, but this case, being so horrific, was hard to forget. His suffering and his death were so utterly pointless.

The following December, Matthew Welsh, nineteen, was convicted of murder and ordered to serve at least twenty years. Nathan Barnett, twenty-seven, was ordered to be detained indefinitely in secure accommodation under the Mental Health Act, after he pleaded guilty to manslaughter on the grounds of diminished responsibility. Sarah Morris, seventeen, was jailed for four years after she was cleared of murder but convicted of deliberately attacking Adam. Daniel Biggs, nineteen, was cleared of murder and inflicting grievous bodily harm but admitted to conspiring to pervert the course of justice and was sentenced to two-and-a-half years in custody. The judge at the sentencing hearing said the gang lacked humanity.

Chapter 11

There was no time to decompress after the Adam Morrell and Laura Torn searches. Almost immediately we were contacted by another police force to join the search for a missing child, suspected of having fallen into a river in the seaside resort of Great Yarmouth.

Daniel Entwistle was seven years old and last seen near his home in the town on Saturday, 3 May 2003. His parents, Paula and David, became concerned that evening when he failed to return for tea. Frantic, they started looking for him, enlisting the assistance of friends and neighbours. Norfolk Police were alerted and a major search operation swung into action.

CCTV from a local convenience store picked up Daniel near his home at around 5.05 p.m. that day. He had also been seen with a group of other boys playing at a quay on the River Yare, which ran through the town to the sea.

In the early hours of the next morning, Daniel's red BMX bike was found abandoned near the river close to the quay. Soon after this ominous discovery, we received the call from the PolSA on the operation, asking if we could search the river for three-foot-two-inch Daniel.

The River Yare is tidal, which meant that if Daniel was in the water, his body was likely to have moved in the previous twelve hours. Success in any tidal underwater search is heavily reliant on how fast the river is flowing, how long the delay was before the search began and whether the body was lodged or trapped under debris below the surface. You can predict how far and in which direction a submerged body will move up to a point if you have an understanding of river currents and the tidal patterns. The composition and size of the body also has a bearing on how it moves, as does the rate of decomposition. Generally, in slow rivers, lakes and ponds bodies will decompose and microbiological breakdown will cause the body to bloat and float to the surface, if it hasn't been caught under any obstructions.

In the open sea, however, there are too many variables to accurately predict where a submerged body may be or might float or wash up. As a rule, the sooner a search gets underway, the fewer variables there are to deal with and therefore the more chance the searcher has of finding the target.

For Daniel, we started our water search within forty-eight hours of his disappearance, which in some circumstances can be enough time to recover a body, but as he was in tidal water, we knew there was a high likelihood that he could already have been washed out to sea. Every six hours there's a high or low tide, which is a big movement of water. In a river that isn't near the sea and has a high and low tide a body will generally stay in a certain area, being washed up and down the riverway depending on the tide. But in an estuary a body will get washed out to sea on the ebbing tide, where it is then very hard to find because there are so many currents that will take it away.

We searched thoroughly but could not find him. All we found was debris that had been dumped by someone renovating an old paddle steamer in the quay.

Sadly, Daniel's fate remains a mystery to this day. One line of inquiry suggested that he may have been the victim of a paedophile. In 2015 his father, who was known to have problems with alcohol dependency, was found dead in his home aged fifty-three. It was later reported that he had served a six-month prison sentence in 1987 for sexual intercourse with a girl under the age of thirteen. Police were aware of the conviction when Daniel went missing and David was questioned but was never a suspect. I was starting to learn that, sadly, despite our best efforts, mysteries often remain unsolved and people remain missing.

The grim roll call of missing persons searches continued. The next call I received was from Andy Baker, SIO in the Met Police's Specialist Crime Directorate. He was working on an active case that was horrific on every level.

Three generations of one family had gone missing from their home in Hounslow, London. Amarjit Chohan disappeared on 13 February. Two days later his wife Nancy and their sons Devinder, eighteen months, and Ravinder, two months, and Nancy's mother Charanjit, also went missing.

Initially there was a suggestion that Amarjit, also known as Anil, who owned a successful freight haulage company near Heathrow Airport, had fled the country with his family after getting involved in some shady business deals. However, suspicions were raised following a tip-off that he signed over full control of his company to an associate in a handwritten letter just before he vanished. The associate was Kenneth Regan, whose friend was Amarjit's business partner, and who had started visiting the firm. Regan was a convicted heroin dealer who became a police informant in return for leniency after being sentenced to twenty years in jail. He hatched a plan to take it over and recruited a friend, William Horncy, to help, and as a full investigation was launched, police began to uncover the sickening plot by the two men to get their hands on Amarjit's successful business and use it to launder drug money.

Amarjit had been lured to a business meeting in Wiltshire by Regan where he thought he was meeting a Dutch buyer for his business, but he was kidnapped instead and taken to Regan's home. There, he was tortured, forced to sign a letter and made to record voice messages for his family telling them not to worry and that he'd be home soon, so they thought he was okay. It is believed he was killed two days later, at which time Regan and Horncy hired a van, drove to the Chohans' family home and murdered Nancy, her mother and the two children before taking the bodies away. The family were buried together on farmland in Devon but dug up after several weeks on 19 April, because the killers feared the remains would be discovered. The bodies were then driven to Dorset and the killers took them out into the English Channel aboard a rickety old motorboat and dumped them in the sea.

The boat was seen by a Dorset Police marine unit and the officers on board remembered commenting on it; the weather was poor that day and the officers thought it was risky for anyone to go to sea in such awful conditions. Ten days after being dumped, Amarjit's body was found floating near Bournemouth Pier by a passer-by. A post-mortem examination found sedatives in his system and restraint marks on his wrists and ankles. In desperation, inside his sock he had hidden a note naming Regan.

I was called to help try to recover the other bodies or any other evidence the suspects had thrown overboard. When I arrived, I was briefed on the case. The investigation was at an advanced stage. Amarjit's body had been found. Detectives knew that the family had been kidnapped and killed. They knew their bodies had been taken to Devon, buried on a farm, dug up again and transported to Dorset. I went to the lay-by where police suspected the bodies had been swapped between vehicles and put on a boat on a trailer, which was driven to a marina and launched. I knew who the suspects were, I knew police had cell site analysis that pinpointed their movements and I saw the CCTV footage of the boat being filled with fuel at a local garage.

In order to narrow down the vast search area, I went back to the Forensic Science Service where the boat used by the men had been taken for forensic examination. There, I looked at the size of the boat, its make and the horsepower of the engine and the size of the fuel can. I also checked the hull to see if there were any barnacles on it – barnacles will create drag and as a result will make a boat burn more fuel. I then worked out the fuel consumption on the engine and used all the data to calculate how far the boat was likely to have sailed on its return journey the day the bodies were disposed of. Even with an approximate range, we still needed to search a very large area and there were so many variables the chances of finding the bodies were slim. However, any possible clue in such a big case that could help nail the evil perpetrators was worth the effort. To narrow down the area further I looked at the sea charts, which show depths and any interesting or unusual features on the seabed. Within the search area there was a patch of seabed used for dumping gravel. It was also a good dumping ground for a body.

My team worked in conjunction with the Dorset Police on the operation and we used their marine unit boat, 'The Alarm', from which to dive. We also used SSS and spent a full month diving off the South Coast looking for the remaining bodies and anything else that might have been tipped overboard by Regan and Horncy.

We started off in the Poole area, moving around the coast. Days were spent meticulously scanning the seabed and diving down to investigate any targets we found. It was long, laborious work. Each night I'd retire to my hotel room and plan for the next day. Once an area was cleared, we crossed it off and moved on to the next. It was very frustrating. I found several targets and each time jumped in and swam to the bottom, only to find rocks.

Unfortunately, not through lack of trying, the odds were against us and the search proved fruitless. Everyone was realistic and we all knew it was like looking for a needle in a haystack – it was a massive task – but even if there was a tiny chance, the job still had to be done. We all went in with our eyes open and there were no recriminations or judgement on ability because we didn't find anything. There were down days but I always had two things in the back of my mind: that I was helping the family and that I was helping the investigation.

Eventually, however, the sea gave up some of its secrets. In July, Nancy's body was found wrapped in a sail bag and caught up in a fishing net off the coast of the Isle of Wight. Her mother's body also washed up on the island four months later. The bodies of the two boys have never been recovered.

In July 2005, Regan and Horncy were convicted of murder at the Old Bailey and sentenced to serve at least twenty-three years in prison. There were several key pieces of evidence which helped jurors piece together what had happened, including the cell site analysis, the CCTV that linked the suspects to the boat and the note in Amarjit's shoe. A co-conspirator, Peter Rees, was found guilty of murdering Amarjit Chohan, but cleared of the other four killings. He was also jailed for life. Rees was Horncy's friend; he had posed as the potential buyer and then guarded Amarjit while Horncy and Regan went off to kill his family. He also dug up the bodies from the Devon farm and helped to dump them at sea.

The Chohan investigation was a big case conducted by the Met's Specialist Crime Directorate. It was the biggest case I had been involved in and I realized that I had a skill set people wanted. I was clearly on the map, in terms of national expertise. It wasn't a proving ground as such, because I was already well experienced and had been involved in numerous investigations, but as a private contractor, to be trusted to work ably side by side with officers for a month showed a high level of capability.

As if to confirm just how valued my services were, soon after the Chohan case, I was asked to assist in a notorious cold case double murder in Wales, following new information that had been received after a television *Crimewatch* appeal.

The bodies of Harry and Megan Tooze, both in their sixties, were discovered in July 1993 at their remote farm in Llanharry. They had been shot in the head with a twelve-bore shotgun. After the execution-style killing, their remains were covered in a piece of carpet and hidden in a cowshed under some hay bales.

A neighbour heard the gunshots and alerted the police, who found their home unlocked and a half-prepared lunch untouched in the kitchen.

Their daughter's partner, Jonathan Jones, was arrested and charged with murder after his fingerprint was found on a china cup at the home. The prosecution claimed he wanted access to the couple's £150,000 inheritance. His defence argued that the fingerprint could have got there after Jonathan went to the farm to help police. He was found guilty and jailed for life in 1995, but the conviction was quashed on appeal in 1996. He was freed at the Court of Appeal after the three judges said they were all 'of the clear view' that the conviction was unsafe. The disputed fingerprint illustrated the importance of forensic awareness at crime scenes. The crime scene was not secured so it could not be established when the fingerprint was left on the cup. Indeed, the appeal heard there was a lack of any forensic evidence to link Mr Jones to the murder. The Toozes were blasted at close range and the killer would have been showered in blood and brain matter. Police took away Mr Jones's clothes, his glasses and even his wash basin to check, but could find nothing. There were several cases in the nineties that fell apart because protocols around the preservation, removal and storage of evidence were not tightly maintained.

On the tenth anniversary of the murder, an appeal was launched which was broadcast on BBC's *Crimewatch*. Subsequently a member of the public came forward with some new information. At the time of the killings, this person's job was to test the water in an old disused mine. The mine had been closed in 1962 and the shaft capped and fitted with a small metal access point that protruded about three feet out of the concrete cap with a hinged steel lid that locked with a padlock. He remembered that around the time of the killings he was carrying out environmental water checks inside the shaft and the lock on the cover over it had been broken off, suggesting someone had accessed it.

In the original investigation the area had not been searched because there was no information suggesting that it had any links to the murder and it was not near the property. It wasn't a fault in the investigation that it was not searched originally. In order to have been considered as an area of interest, police would have needed specific intelligence, which they didn't have.

I was contacted by the chief inspector, Trevor Evans, who had been given my details by the National Crime and Operations Faculty, who explained that the access shaft that he wanted to be searched was around 750 feet deep, and to make matters worse, it was flooded. Although it sounded like a difficult search, I knew I could do it. The fact that the shaft was full of water would make it easier to search with some equipment I had recently invested in from the US, which included a newer, smaller ROV and a drop-camera with a 1,000-foot cable.

I was tasked with searching for anything related to the murder, in particular a shotgun, the shotgun cartridges and blood-stained clothing. I lowered the ROV down fifty feet into the shaft through the access point before it hit water. The team paid out the cable and I watched the video monitor as the ROV descended through the eerie darkness, its four powerful halogen lights creating shadows on the shaft wall.

As the ROV came within three feet of the end of the shaft, I slowed it down to rest gently on the bottom, preventing a big silt cloud from kicking up that would hamper visibility. We then lowered the vertical-drop camera and positioned it a few feet above the ROV so I had a clear view of it and the area around it.

The bottom of the shaft was covered in decades of silt which needed to be carefully swept away before any secrets beneath it could be revealed. If there was something down there, I was confident that I would find it. I sat in our command trailer in front of two video monitors, one to view the ROV camera and one to view the drop camera. The drop camera cable and the ROV cables were running parallel to each other over two pulleys hanging beneath a tripod. I took a sip of tea and worked in silence. I couldn't see anything on the drop camera or the ROV camera that stood out so I turned on the ROV motors and carefully flew it around the bottom of the square mineshaft – the pictures were crystal clear. I searched every nook and cranny but saw nothing remarkable. I then manoeuvred closer to the silt and, realizing that I could use the rotors of the machine to gently 'sweep' the silt, I used the electric propulsion motors to fan the silt out of the way. I did this section by section, allowing the silt clouds to settle before moving on to the next area. It was a time-consuming and delicate search, but after hours of searching I saw something that grabbed my attention.

'I've found something,' I said to Trevor. 'It's a shotgun cartridge, it looks like a twelve-bore.'

Trevor just stared at me, speechless.

'Does the cartridge used in the murder have any identifiable markings on it?' I asked.

'The word "game" should be written on the cartridge casing,' he confirmed.

I manoeuvred the ROV closer to the cartridge, zoomed in, and sure enough, there, on the side was the word 'GAME'.

It was a perfect find and Trevor was ecstatic. He was in no doubt it was the cartridge that had been used in the murder. On the strength of the find, I was asked to come back and conduct a week-long search of the entire mineshaft. During the meticulous operation I also found important items of clothing and a holdall that had been dropped down through the open lid into the flooded shaft. The two items had floated down and landed on two different support beams.

We worked in collaboration with the South Wales Police underwater search dive team who had to access the shaft via a horizontal passage and were able to successfully recover the evidence.

While I was working on the search, we started to attract the interest of some of the local children, a couple of whom asked what we were looking for. One of the important lessons I had learned about searches was that you should always listen to local people as they know the area better than you do, and they also often have important information. So I had a chat with the kids, assuming that they had probably been playing in the area for years and might know of other hidden places worth searching. I told them what we were looking for and asked if any of them had ever heard of anyone finding a shotgun or cartridges in the area.

'My brother found shotgun barrels a few weeks ago,' one of them shouted out. I raised my eyebrows. I called Trevor over.

'Really? Where?' I asked.

'In the brambles over there,' he pointed.

I asked him where the barrels were.

'My brother took them home, painted them pink and hung them on his bedroom wall.'

One of the officers I was with, who had first ignored the kids, was suddenly very interested.

'Can we come back with you and have a look?' he asked.

The barrels were recovered and sent off to the ballistics testing lab in Chepstow. I didn't hear back whether they were a match or not, but I suspect they were. On most police jobs, once my work had finished the investigation moved on and I went to other jobs.

Despite my significant finds, the case remained a mystery and the murderer has yet to be brought to justice. If there is a trial in future, I've no doubt the items I recovered will contribute to the case.

In the same year, my team and I were engaged in searches for bodies and evidence in other well-known cold cases, in the hope that our experience and equipment would lead to breakthroughs.

The first of these was the case of Colin White, an engineer who went missing in February 2002. His wife, Anne Dickens, was initially a suspect and was arrested in 2004 but never charged. Another suspect was a builder. He also was never charged. We were called in to search the couple's recently renovated house with GPR. We were primarily looking for a body.

Our first target area was the garden, which was well maintained. Since Colin's disappearance there had been a considerable amount of renovation work on the garden and house. In numerous histor-ical murder cases, culprits have used building work and garden landscaping to hide the fact that they are depositing their victims. Some killers like to keep the remains of their victims close, as was the case with serial killer Fred West. We scanned the garden and found nothing so we moved into the kitchen, which had also been refitted. Had someone been looking to entomb Colin's remains somewhere close but inaccessible, a grave under a new concrete floor on top of which kitchen units were fitted would have been an ideal opportunity. Because of this I needed to search under the units, right up to the walls, and so unfortunately had to insist that the units were removed, which they were thanks to the over-enthusiastic efforts of the police tactical support group. They were not as delicate in dismantling the kitchen as I'd hoped. When we cleared the kitchen and could confirm there was no trace of a body there, we searched under the floorboards in the rest of the house as cavities under floors also present opportune places to hide bodies.

After a very thorough search I was confident that the house was clear. We never found any sign of Mr White, who is still missing at the time of writing and the case remains a mystery.

Another case in which the GPR was used was that of missing teenager Sarah Benford, who was last seen alive in April 2000. The troubled fourteen-year-old was a drug user and evidence suggested she was trafficked to London for sex with older men. There were sightings of her for a couple of months after she went missing but the case ran cold and by 2003 police announced they were certain she had been killed.

I was called in 2004 to run GPR searches at a house connected with the investigation.

The property was in a horrendous state, full of overflowing ashtrays and filthy stained furniture. The condition of pet ducks kept at the bottom of the field attached to the property was so bad I called the RSPCA, and when they couldn't take the animals, we cleaned the rancid cages ourselves and provided fresh straw and food.

In the garden just outside the back door the GPR showed a large anomaly under the patio that could have been a body. I reported this to the PolSA, who told me that it was a pipe that didn't need further investigation. I asked for the plans for the house to confirm the information but was told not to worry, and that the job was finished.

I was not satisfied, however, and rang the national search

adviser to voice my concern. The National Crime and Operations Faculty were informed but the report was never acted on. The loose end bugged me for years and I refused to let it go.

In my view, you did as thorough a job as you possibly could when you searched for both evidence and bodies and left no stone unturned. Ultimately, my job involved bringing closure. The knock-on effect of a missed clue or a premature end to a search could be years of uncertainty for families and friends who just want to know what happened to their loved ones and where they were.

Ten years later I was doing a forensic presentation to a group of crime scene managers when an officer from Northampton-shire Police came to speak to me and told me he was reviewing the Sarah Benford case. I told him about my concerns, and he explained that he wasn't aware of the anomaly. I sent them the original file, including the letters that were sent to the force. I had a follow-up meeting with the SIO and presented my findings. I told them that the anomaly must be checked and I offered to re-run the GPR over the patio again. Months went by and nothing happened. I rang the crime scene manager, who advised that the area had been searched again by the Home Office and reported there was nothing there. I still wasn't convinced and a few months later the investigative journalist Mark Williams-Thomas approached me as he was making a documentary about the case. He went back to the address, spoke to the woman who lived there and got permission to go back and search properly.

Over a decade later I took the GPR back and found the exact same anomaly in the same place. It certainly wasn't a pipe. Chris and another member of my team, Steve, dug down a few feet

Searching serial killer Peter Tobin's garden in Bathgate using ground-penetrating radar (GPR), with Chris and Robin.

Using GPR to search Tobin's kitchen.

Conducting searches at Tobin's former address in Southsea, Hampshire. No longer the men in black, more like the men in white.

Keeping a sharp eye on the roof of the Palace of
Westminster during Pope Benedict XVI's 2010 visit.

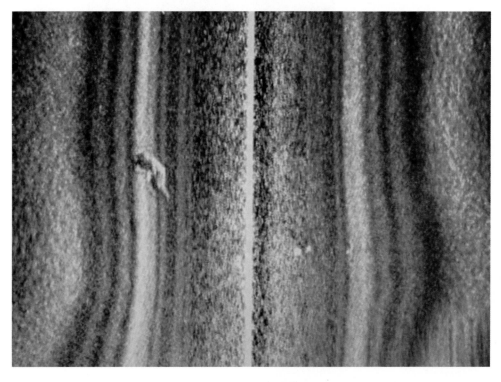

A drowning victim as seen on sonar.

Preparing to dive on Damian Tudge's car.

Forcefully opening the door of the car to allow us to remove his body.

Me finding three buried handguns during the
Hieronim 'Henry' Jachimowicz case.

Looking into Henry's grave under the path in his own garden.

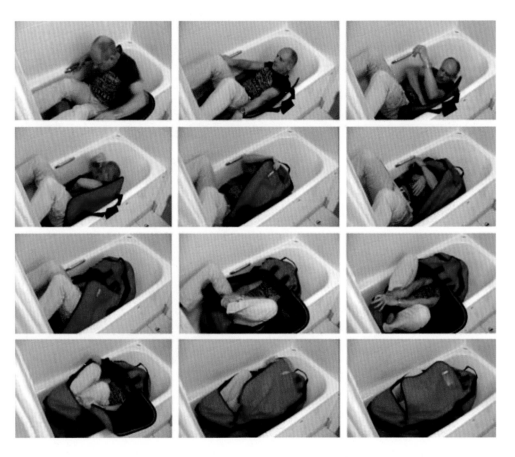

Me carrying out experiments to see if it was possible for MI6 officer Gareth Williams to have padlocked himself into a bag in the bath.

Working around the clock with my team on the 2013–14 floods in Surrey.

Searching a pond full of weeds during the Headley pool party murder investigation.

Retrieving a vital piece of evidence for Scott Wilkinson's murder.

Using GPR to search for Nicola Payne, who has been missing since 1991.

Searching deep air shafts for Linda Razzell.

With my daughters Danielle (*left*), Summer (*centre*) and Natasha (*right*).

Adele and I on our wedding day in Canada, August 2012.

below the surface and found a large sinkhole about three feet in diameter and full of gravel – a perfect disposition site for a body. It was apparent that the Home Office team had not searched as thoroughly as they could have. We removed the gravel, and although Sarah wasn't there, and had never been there, and the excavation gave no clues as to what happened to her, my tenacity did finally close off one loose end.

Not every job provided a result. It would be great to be able to write how in every case I was called to I managed to uncover a vital piece of evidence, or a body. But that's not real life. Crime is a tangled and confusing business and it's the police's work to try to untangle it all and piece together all the loose ends. Often leads go nowhere and sometimes evidence and victims are never found. This doesn't mean you stop looking, though, or you discount a lead that seems unlikely. Every stone needs to be overturned, every hunch followed and every piece of information followed up on. I made sure I went into every job with an open mind and a commitment to see it through for as long as it took, until I could be confident that I'd searched everywhere. Of course it was disappointing when I couldn't find anything, but at least I could console myself with the knowledge that I didn't find anything because there was nothing there, not because I hadn't looked hard enough.

Often, the range of call-outs we were engaged in swung from the sublime to the ridiculous. I was highly regarded, and my work was appreciated. I had skills that were valuable. I could search under water, under buildings, in drains and tunnels, I could climb buildings, look in gutters. I was useful, a problem solver, a search Swiss Army knife.

Soon after the case-breaking searches in the Tooze murders,

I was called to what appeared to be a more run-of-the-mill job. A contact from Avon and Somerset Police made the call.

'Peter, we need you down here. We've got an escaped convict in a tree. He's armed with a knife and he's refusing to come down.'

Within a few minutes I found myself in one of SGI's emergency response vehicles, heading towards Shepton Mallet. On the way I got more information. The man was a convicted paedophile who had escaped from the sex offenders' unit he was in and was heading to see his wife. His wife rang the police and he was chased into a cemetery, where he had clambered up a tree around which a crowd gathered, hoping for a lynching no doubt. He was armed and was threatening to harm himself and anyone who tried to get him.

Avon and Somerset Police didn't have a rope team at that time, so they called me. By the time I got to the scene in the late afternoon the man had been in the tree for the best part of a day. Two police negotiators had been talking to him for hours but had failed to make any headway. I got the impression that the guys were not that pleased to see me, probably due to their professional pride taking a dent by having an outsider brought in. After an awkward initial exchange in which I was told that civilians were not allowed to conduct police negotiations, I managed to persuade them that I knew what I was doing and I got my gear on, put some food and drink in a backpack, borrowed a stethoscope from a paramedic at the scene and started to scale the tall fir tree the escaped con was in.

'I'm a medic,' I shouted up to him, hoping the little white lie would make him more amenable to me. 'I'm coming up to make sure you are okay. Don't worry, I'm not a police officer.'

I was wearing a rescue suit.

The man was right near the top of the tree, nestled in the crook of a branch. He was skinny and balding and looked tired. As I approached, he tried to scare me off half-heartedly.

'Don't come any nearer, I've got a knife,' he called. He waved a Swiss Army knife towards me. It was obvious he wasn't going to use it, unless I had a tin of beans on me that he wanted to open.

'Just put your knife away,' I said to him. 'You're only going to make a bad situation worse. No one's going to hurt you. I just need to check your vital signs and make sure you're okay.'

He calmed down and let me get near enough so I could go through the motions of pretending to be a medic and check his pulse.

'You're dehydrated,' I told him. 'You need to drink something.'

I pulled one of the drinks from my bag and handed it to him. He drank it thankfully.

Even though I knew the man was a sex offender and had no sympathy for him, I was there to do a job and so I put any feelings I had about him to the back of my mind and started to talk to him, calm him down and introduce the idea that he needed to start considering climbing back down.

'You're not achieving anything and the longer you stay up here, the more trouble you're causing yourself,' I explained.

Just as I was establishing a rapport, I heard a commotion below and we both looked through the branches.

The crowd that had surrounded him earlier had dispersed and gone to a nearby pub, but it was 11 p.m. and chucking-out time and they were back at the cordoned-off entrance to the cemetery, talking to the officers there and shouting abuse at the man. One very large man was being particularly vocal.

'Let me through, I'll get him down,' I heard him say.

And then I watched as he trotted round the perimeter of the walled graveyard to right underneath the tree we were in. He then climbed onto the wall, which was around six feet high, and launched himself at the lower branches. For a big man it was quite impressive. Then he started climbing towards us, sending judders up the trunk as he did so.

The escaped prisoner started to panic.

'I'm going to jump,' he screamed.

'No, you're not,' I said.

One of the officers ran over and grabbed the overweight vigilante's leg, pulled him off and led him away for a ticking off.

After the tree-hugger calmed down, one of my other team members swapped places with me to give me a break. Eventually, within three hours of us arriving at the scene, the man was persuaded to climb down. I used the same communication techniques with him as I had previously with the protesters when I needed to establish a rapport and gain their trust. I put my feelings aside and just chatted with him about his life, why he was there and how he was feeling. I made him feel safe and he realized that I was not there to judge, just to help resolve the situation safely. In all the jobs I did I never saw it as my role to judge – that was a job for the courts and the justice system.

As I left, the police negotiator thanked me and commented on the excellent job. I never heard what happened to the man after that, but likely he would have gone back to prison.

A few weeks later we were back to a completely different job, a world away from prison escapees, dismembered bodies and hunts for evidence, when the rope team was utilized by the British

Transport Police to help deal with protester activity. This time it was on the London transport system during a big defence industry conference called DSEI (Defence & Security Equipment International) at the Excel Exhibition Centre, in Docklands. It seemed that one minute I was wearing a forensic suit running a radar over a suspect's garden, the next I was in a diving suit trawling the SSS through a lake looking for a body, and the next I was in a rescue suit coaxing a protester from a building.

In this job activists against the arms trade were planning to stop people getting to and from the event by sabotaging the Docklands Light Railway (DLR). Prior to the operation my team and I received safety training on electrified railway tracks, to make sure we didn't fry ourselves in the process. During the course, which took place at a London Underground interchange in Acton, we were shown CCTV footage of one poor chap who tried to steal copper cable from a railway line and accidentally touched the electrified rail with bolt cutters. The voltage went through his ankle and out of his head, which erupted like a volcano. It certainly had the desired effect of focusing the mind.

On the day of the conference we had a morning briefing with Special Branch in which we were told that there wasn't predicted to be much activity. We were sent to a holding area on a patch of wasteland near Docklands in a place called Silvertown. We were there with contingents from the British Transport Police, London Fire Brigade and the Ambulance Service. The first incident came early in the morning as rush hour was getting underway.

A young protester was D-locked to the front of a DLR train. When we arrived, a crowd of angry commuters had gathered around him, some of whom were threatening to take the law into their own hands and remove him themselves, along with parts of his anatomy. The poor guy looked terrified and was more than happy to let me remove his lock and lead him away.

We then got called to another guy who was at a different station, wearing a rucksack, sweating and pacing up and down on a train roof. When we'd established that he wasn't a terrorist, I climbed onto the roof with him and talked him down.

The protests continued throughout the day. The next call was three girls on top of another train, two of whom were D-locked together. We used hydraulic cutters to separate them, then we put them in stretchers and lowered them to the platform, where they were arrested. One continued to run along the train roof, so I put a distraction plan in place. I went from one end of the train to the other with three team members carrying assault ladders. While the protester was being distracted the rail engineer carefully opened the doors and we quietly placed the ladders at each end of the carriage. Another team was ready to get onto the tracks to catch her in case she fell. On the command, two of us silently climbed the ladders onto the roof and before she could flinch, we had hold of her. She was carefully strapped into a rescue stretcher and lowered to the ground.

Then, at another station, a protester turned violent when one of my team went up on the glass roof to remove him. He needed to be restrained and handcuffed for everyone's safety. There was also a violent reaction from one of two female protesters during the last job of the day. One was on a train roof and the other jumped from the train to the gantry of the station. The girl on the train punched and kicked when we tried to get her down, while the one on the gantry, realizing there was a sheer drop on the other side of the roof, was more compliant and came down easily.

The protesters we dealt with at that job seemed like a different breed from the eco-warriors of the nineties. For one thing, they were quiet. There was none of the community and the banter. They didn't know us and it looked to me as though they didn't know each other. That, I assumed, was the result of the internet and social media. Individuals could now organize and come together easily without having ever met, and whereas in the past protesters had organized themselves into close-knit networks, now strangers could find out where and when a protest was happening and get involved. Pick a cause, turn up. It was that easy.

It made the job harder for the police, of course, as it was a much more informal way of organizing a protest. There were no communities to infiltrate. It explained why, at the morning briefing, we were told not to expect much activity – there simply wasn't enough intel to suggest otherwise.

Chapter 12

As the ten-year anniversary of the first big job at Newbury approached, my life became a constant whirr of emergency rescues, protester removals, crime scene investigations and body recoveries.

And it was the latter that always had a lasting effect on me. Human remains tell stories about the death of the person they belong to. Sometimes they are at peace, lying in a bed having passed away in their sleep at the end of a long, fruitful and loving life, but these were not the types of stories told by the remains of the people I was asked to search for.

From Adam Morrell's dismembered body parts to that first boater in the lake gripping the reeds that kept him anchored underwater until his lungs finally filled with water, these remains spoke of tragedy and lives cut short.

As the years progressed, such stories continued.

In April 2004 I was asked to investigate a particularly sad tragedy after two divers disappeared, presumed drowned, at the National Diving and Activity Centre in Gloucestershire over the Easter holidays.

Hank Austin, a chef, was twenty-five and had gone to the centre, based at Britain's deepest flooded quarry, with his sweetheart, Janine Davison-Evans. Both had completed diving courses and Hank was a more experienced diver, having completed ninety-three different dives. The couple had only been together for four months and had met through their shared passion for the sport at a Christmas party held by the Bracknell Dive Crew in December 2003.

They were reported missing after staff at the dive centre noticed Hank's car was still in the car park as they closed for the night. It was obvious that something terrible had happened and the South Wales Police and Avon and Somerset Police underwater search unit were called in. Soon after, I was contacted by Sergeant Bob Randall, head of the unit, who later came to work with me and my dive team. I set off with the team, the ROV and the SSS, ready to face whatever was waiting for us below the surface. On the journey to the job I reflected on my own dive training all those years before in Scotland, and I thought about just how dangerous diving can be if you are not properly trained or equipped. There are lots of variables that can turn a leisure activity into a disaster.

When we arrived, we set to work and initially started looking for the couple using sonar. The quarry went down at different levels, from twenty-five metres down to fifty and then to seventy. Each descent was marked by a ledge, and they were covered in large boulders. If the bodies were between these rocks the sonar would not find them, so instead we used the ROV and sat for hours flying it between the huge rocks below. We searched late into the night. I flew the machine down to seventy metres for what was going to be one of the last passes of the night and Darren, one of the team, took over while I went to brief the SIO.

As I did, Darren called out.

'I've got a target.'

I ran back and looked at the screen, on which was the unmistakable image of two people. They were lying together between large boulders at the seventy-metre level, clipped together with a buddy line, one on top of the other, motionless in the still water. They looked peaceful.

It was too late in the night to recover them. That job would have to wait until the morning. They had been dead for hours and there was no point risking another life sending someone down so deep late at night. Instead, to enable us to find them again in the morning, I secured the control handle of the ROV with an elastic band to hold it hovering in place over them like a sentinel and switched off its lights, because it seemed like the right thing to do to give them some dignity in their resting place.

The following morning, I turned the lights on, and they were there on the screen, exactly as they'd been left.

The depth the bodies were at presented us with a problem. It was beyond the limits of the police dive team, who are limited to fifty metres, the same as my team. As we were discussing how to recover the bodies one of the instructors at the centre came over and asked if he could help. The depth requires a 'trimix' diver. Trimix is the mix of three gases – oxygen, helium and nitrogen – that divers need at greater depths to help them breathe safely. The instructor explained that he was a qualified trimix diver and volunteered to go down and attach a line to the couple so we could pull them up to a depth where we could then recover them.

We were not allowed to instruct him – it had to be his choice and his choice only. We explained it to him and he was happy to dive. He went away and prepared his equipment, and an hour later he started his slow, eerie descent into the darkness.

After several minutes, I saw him swim into view on the ROV

camera and secure our line onto the kit of the missing divers. It must have been a distressing sight for him, as most people are thankfully not used to seeing dead bodies. He swam away quickly, completely spooked by what he'd seen.

A plan was put in place using two boats. Bob descended to fifty metres while we pulled the pair up to him. Bob then secured another line onto their harnesses. With Bob and the bodies secured on a line attached to the boat, I then drove across the flooded quarry to land the divers on a shelf at fifteen metres. At that point Bob needed to decompress so another police diver went in to meet him and untangle the two bodies. It was a delicate and slow task. The couple's diving equipment had to be removed underwater, except for their masks, and then their bodies were placed on a stretcher and carefully hauled to the surface. Once out of the water we were careful not to move their masks as any residual fluid in them could be tested to help pathologists work out how they had died. The bodies were taken to the forensic tent. It was another waste of life.

The dive centre kindly laid some food on in the cafe and we sat outside with the police divers. A lady walked across the car park and said she was the mother of one of the victims. She was crying and wanted to thank us all for our efforts in bringing them back. All the grown men around the table went silent. Several blinked back tears. No more words were spoken; it was extremely distressing for all involved.

Hank's wrist dive computer gave some very useful information. It showed that the couple had gone on two dives that day and, on the second, a rapid descent to the bottom was recorded. Hank's cylinder had no air left in it, but Janine's did, suggesting she died before her air ran out.

The previous week, Hank had been diving in sea water, which requires extra weight to counteract the buoyancy of the salt water. When we found him, he still had extra weight attached to his diving belt, which led me to wonder whether he sank from the extra weight and took Janine down with him.

An inquest conducted in 2005 found that Janine had died from 'barotrauma', a condition which causes the lungs to tear due to pressure. Hank drowned. It was never fully ascertained what happened to cause such a catastrophe, whether one pulled the other down with them to a watery grave, but the sight of the two lovers, bound together in the dark water, stayed with me for a long time.

Chapter 13

History usually judges victors more kindly than those they defeat, but the crew of Greenpeace climbers who descended the 750-foot chimney of Kingsnorth power station in October 2007 were, in hindsight, on the right side of events. They may have lost the battle then, but they have certainly won the war now.

The audacious stunt started just before dawn when six climbers from the organization managed to infiltrate the coal-fired station and scale one of the chimneys, forcing the whole plant to go offline. They were protesting the use of polluting fossil fuels for generating electricity – well ahead of the curve.

We were called to the power station urgently on the day of the protest to issue them with an eviction writ on behalf of the bailiffs. When we arrived it became immediately apparent how the team had managed to trespass. Putting it kindly, the security was relaxed. Some of the perimeter fences had fallen, allowing intruders to walk straight in.

As an organization Greenpeace was growing in authority and reputation. Only a few years previously it came close to being a proscribed terrorist organization in the US, but it was vindicated after a controversial FBI investigation into it and other groups, including PETA, was discredited. By the time the protesters audaciously scaled the brick chimney stack in Kent, Greenpeace was renowned for doing valuable work and raising awareness of issues such as rainforest loss and whaling.

Its direct-action stunts were usually spectacular and were always carried out professionally and safely. I'd encountered the group before and knew that it used professional, experienced climbers. I had respect for the people and the organization. In terms of protest activity, Greenpeace was the gold standard. It chose its activities carefully to make maximum impact.

I knew as I drove to the power station in Kent that I was unlikely to have any problems with violent behaviour, either, and that the only issue I was likely to face was difficulty getting to them because they used good climbers who always got to the most tricky places.

I needn't have worried, though, as when I got there they had already come down of their own accord, having painted the word 'Gordon' vertically on the huge stack, in reference to the prime minister of the time, Gordon Brown. I thought it just needed the word 'gin' added, to make it into an advert that could be seen for miles. Our job was to go up to check the area where they had staged their protest to make sure it was clear and safe.

The chimney consisted of four individual narrow stacks enclosed in an outer shell that acted as a windshield. The route to the top was inside the outer shell shield. Two of the four chimneys were still hot, so the heat inside on the long climb up was phenomenal. Added to that discomfort was the fact that dust masks also had to be worn because the interior was full of black soot.

We made our way up via the internal ladders, pausing to rest on the ledges at every hundred feet. I carried a bag of water on my back to stay hydrated. It took an hour to reach the top and step outside onto the outer platform from which the protesters had launched their painting expedition. From there the view was breathtaking, and in the wind I could feel the whole structure swaying.

At the time the job seemed like a run-of-the-mill search and secure operation, but afterwards the protest became a watershed moment in the history of environmental activism. The six eco-warriors were charged with causing an estimated £30,000 of damage to the power station and went to trial in Maidstone Crown Court in September 2008. They admitted trying to shut the station down but argued that they were legally justified in doing so because they were trying to prevent climate change causing greater damage to property elsewhere around the world. They presented evidence from expert witnesses including climate scientist James E. Hansen and an Inuit leader from Greenland, who all confirmed that climate change was already seriously affecting life around the world. A jury cleared all six in what was thought to be the first case in which preventing property damage caused by climate change had been used as part of a 'lawful excuse' defence in court. The defence has been used in subsequent cases.

A month after the trial, protesters occupied part of the site again in protest against owner EON's plans to build two new coal-fired units there. The tide of opinion and policy was turning away from coal and in 2010 the company announced it was withdrawing the plans. The power station, which had not been adapted to meet emissions targets, closed in 2013 when it ran out of its allocated operating hours under EU environmental laws.

In the same year as the Kingsnorth job I was called to another sensitive protest, which also had repercussions long after the last protester was removed.

The call came from the Mayor's office in City Hall, London. An activist's camp had been set up in Parliament Square opposite the Palace of Westminster. It had been dubbed the London Peace Camp and was established in protest at the war in Iraq. Earlier in the year a huge march had taken place in opposition to the conflict and feelings were still running high.

In the weeks prior to our operation in August 2007 around thirty tents had been erected across the grass area of Parliament Square Garden. Officials said the majority of these were unrelated to any authorized protest. Complaints had been made by neighbours, staff and MPs and the famous square had turned into an unsightly public health risk, full of litter and human waste. One protester had set herself up on the empty plinth, which was due to display the statue of former prime minister David Lloyd George. The statue had been taken off the plinth for cleaning.

The powers-that-be were anxious to remove the protester that day.

'We need them removed at 6 p.m. tonight so we can get it on the live news bulletins,' my contact at County Hall said. 'I need a quote for the work as soon as possible.'

'That's not quite the way it works,' I explained, and said that before we agreed to anything we needed to discuss operational requirements. I caught the train to London and headed for a meeting at County Hall while the team prepared for the operation. At the meeting the Met Police commander recommended to the Mayor's representatives that they take my advice and carry out the operation after 10 p.m. under the cover of darkness. After some negotiations and my insistence that we were not prepared to turn the operation into a media circus, a contract was signed.

My team assembled later that evening and I briefed them

before getting changed into protective coveralls, harness and helmet. My target was the tent on the plinth, which had become a symbolic focus. As the operation started I made a beeline for it with a ladder, which I placed against the side of the structure and climbed. I looked inside the tent.

'Excuse me,' I said to the middle-aged lady inside.

'Go away,' she ordered. 'You're not dragging me off here.'

'Nobody's going to drag you off anywhere,' I assured her. 'I just want to talk to you. Would you like a coffee?'

By now it was the oldest trick in the book.

'I'd love a coffee,' she said. 'I'm cold.'

I got her a hot drink, asked her if I could sit with her and sat down in the tent for a cuppa and a chat. She said she was a mother and she had been at the camp for several days.

After a while I levelled with her.

'Look, at the end of the day we're going to have to get you down. I know your friends have given us lots of abuse but we're just doing a job, simple as that. We've been instructed by the Mayor to get you down and make sure your welfare is okay.'

Outside someone was hurling abuse up at me.

'Be quiet,' she screamed out. 'He's a nice man.'

I asked her where she was living, and she said the tent was her home.

'I tell you what,' I said, 'it's not very comfortable here, so if you come down, I'll see if I can get you a sleeping bag and a soft mat.'

'Really?' she asked.

'I'll see what I can do.'

On the promise of a new sleeping bag she allowed me to strap her into a stretcher and we lowered her to the ground.

As I unstrapped her another protester sauntered up to me aggressively, hurled a load of abuse at me and said: 'I hope your mother's proud of you.'

Quick as a flash the lady stood up and slapped him straight around the face.

'Leave him alone. He's one of the good guys,' she scolded.

That night, when the site had been cleared, we put barbed wire up on top of the plinth to protect it from any more intruders. The whole area was fenced off to enable clean-up and maintenance to take place and prevent further unauthorized camping.

But that wasn't the end of the Parliament Square Camp. We were back again in July 2010, three months after peace protesters moved in again and established 'Democracy Village'. They fought through the courts to be allowed to stay and the Court of Appeal finally rejected their application, allowing bailiffs and SGI to remove them.

This time the protesters had built some rickety wood and scaffolding structures in the camp that they could lock on to and obstruct the operation. As before, the camp was a seething health hazard and campers had been urinating and defecating in the surrounding bushes, in which they'd also dumped their rubbish and their empties.

The key to that operation was seizing a tripod that they'd set up in the middle of the camp before anybody climbed onto it and locked on. For that reason, we used the element of surprise, hid all the vehicles in side streets and went in swiftly at 1 a.m. One lady managed to lock herself to the scaffold structure but I had a chat with her, gave her a cup of coffee and then carefully cut the lock and removed it.

Another protester climbed onto an eight-foot-high wooden

platform and was being vocal and aggressive. I got the ladder up and climbed up to check on his welfare. Suddenly he grabbed me and bit my arm. Luckily, I was wearing a Kevlar sleeve. As we were tussling, the whole platform gave way. We both dropped and he landed on top of me. The team managed to get him off me and he was taken away by bailiffs, and I was uninjured.

Once again, the site was cleared and a fence was put up around the square to stop anyone returning. A group continued their protest on the pavement, with some brandishing banners emblazoned with the words 'the dispossessed'. They threatened to go back when the fence came down.

The eviction orders and these shenanigans never applied to the best-known protester in Parliament Square, however. Brian Haw started camping there in 2001 in protest at UK and US foreign policy. He successfully fought to be allowed to continue his demonstration after it was made illegal to hold an unauthorized protest within a square mile of Parliament in 2007. He claimed that the law did not apply to him because his protest started before the legislation was drafted. He continued his protest and legal fight for almost ten years and only left the site in 2011 due to ill health. He died of lung cancer while receiving treatment in Germany in June 2011.

Chapter 14

In 2006 and 2007 I became involved in two of the biggest murder cases of the decade, both of which required myself and the team to put our feelings aside and maintain an objective, professional detachment from the abject horror of the crimes we had been called in to investigate.

On 25 January 2006, Rahmat Sulemani called police to report his girlfriend missing. Although it had only been a day since he'd spoken to Banaz Mahmod, he had good reason to be concerned.

Banaz was from a strict, traditional Kurdish family who had claimed asylum in 1995 when Banaz was ten years old. Her parents, Mahmod Babakir and Behya, disapproved of their daughters mixing with people from outside their community and forbade them to wear western clothes. One of Banaz's sisters, Bekhal, was beaten and threatened for having non-Kurdish friends and wearing western hairstyles. She eventually fled the family home in Wimbledon and went into hiding after her father threatened to kill her and the rest of the females in the family. Some within the community saw her father's failure to control his daughter as a sign of weakness.

Two of her sisters had had arranged marriages – one at the age of sixteen with a man fifteen years her senior – and when Banaz was between sixteen and seventeen she was forced into an abusive marriage with a much older man from the remote village in Iraq that her family came from. She was deeply unhappy. In recordings she made she detailed how she had been raped and beaten on many occasions. She also reported the abuse to the police. When she told her parents she wanted a divorce, they replied that divorce would bring further shame on the family and that this would reflect on the wider community.

After two years of marriage, she could take no more, however, and left her husband, returning to the family home in Wimbledon. Banaz's father and uncle, Ari Agha Mahmod, disapproved and were further angered when she began a relationship with Rahmat, who was not from their community.

In a family meeting in December 2005, they agreed she should be killed for bringing shame on the family, in what's known as an honour killing. Banaz overheard a phone call where the plot was being discussed and she delivered a letter to Wimbledon Police Station in which she named five men involved in the plot, who were her father, uncle and three of her cousins.

On New Year's Eve, Banaz's father forced her to drink alcohol and tried to kill her. She escaped through a smashed window and fled to a local cafe, where police were called. The PC who attended did not believe her.

On 22 January 2006, three of the men named in Banaz's letter tried to kidnap Rahmat. He escaped and went to the police along with Banaz. She was scheduled to return to the station two days later but never arrived. On that morning her parents left her asleep in the lounge and went to take their youngest daughter to school. While they were gone three men recruited by her father and uncle – Mohamad Marid Hama, Mohammed Saleh Ali and Omar Hussain – went to the house and subjected her to over two hours of torture and sexual abuse before she was strangled with a ligature. Her body was stuffed in a suitcase and taken away to be dumped.

Initially her parents explained her disappearance by saying she often stayed away overnight, but when Rahmat persisted they were interviewed along with Ari and their homes were searched. Inconsistencies began to surface, and the investigation was taken over by the Met's Homicide and Serious Crime Command. It was led by Detective Chief Inspector Caroline Goode, who contacted me after the suspects had been arrested and one of them, Hama (who Rahmat had identified as one of the men who tried to kidnap him), had been charged and was on remand. While in custody he had been covertly recorded on the phone bragging about the murder and the disposal of the body and Caroline wanted us to search several locations of interest and properties belonging to those involved, primarily to try to find her body.

The suspects had spoken in a Kurdish dialect and one of the transcripts suggested that water had been mentioned when the men were discussing disposing of the body. We were given a list of places of interest to search, which included a funeral director's that the uncle had been refurbishing, a garden of one of the suspects and several drains and waterways.

I used the GPR extensively and checked a casket stored in the funeral director's we were sent to. We checked the drains in all the properties with cameras and searched under buried pipes, because they can provide a good space to hide a body.

We were on the case for a week and there was an unsettling undercurrent to the whole operation, a general feeling that I couldn't put a finger on but that felt like we were unwelcome.

At one point we were tasked with searching a river near the home and, as we were organizing our operations on the bank, people came down and stood on a bridge overlooking where we were and started shouting abuse at us.

We knew about the history of the case at the time, and we knew how tense the situation was between the police and the family, who had pulled rank and were being as uncooperative as possible, so we weren't expecting a welcome party. We did our best to ignore the abuse and nearby we recovered a meat cleaver from the river, which was bagged up and tested but turned out not to be related to the case.

After extensive searches underwater and on land, nothing was found.

As we searched, the detectives on the case continued to press forward in the investigation and one Saturday night Caroline called me to say their inquiries had led them to a garden in Handsworth, in Birmingham, where they had discovered an area of disturbed, freshly dug soil. She asked if I'd take the GPR to the location urgently. As the disturbance was plainly evident, however, my services would not have added anything. They were certain they were at the disposition site, so I told Caroline she didn't need me for this one.

'You need a forensic archaeologist,' I said.

Forensic archaeologists use archaeological techniques to recover evidence and human remains from graves, and I'd worked with them for many years.

Soon after, Banaz's body, still in the suitcase, was recovered from the site in the garden.

During the subsequent trials Hama argued that he played no part in the murder and only helped to dispose of the body. But the jury heard harrowing details of a secretly taped prison visit in which he told an unnamed visitor: 'Her soul wouldn't leave the body. It took half an hour. I was kicking and stamping on her neck to get her soul out.' He described how he stood with one foot on Banaz's back as another man prepared the ligature that would kill her, how he would 'shut her up quickly' and how she had vomited during her ordeal. It was a truly evil crime.

It took several more years for all the men to finally face justice, as after the murder Mohammed Saleh Ali and Omar Hussain fled to Iraqi Kurdistan. In June 2007, Banaz's father and uncle were unanimously found guilty of murder and sentenced to life in prison, with minimum terms of twenty and twenty-three years respectively. Mohamad Hama pleaded guilty to murder shortly after the start of the trial and was sentenced to life, with a minimum term of seventeen years.

In October 2007, Ali was arrested in Iraq after he killed a teenage boy in a hit-and-run incident. He was extradited back to the UK in June 2009, in the first ever extradition from Iraq to the United Kingdom.

Hussain hid out in a remote region but was shot in the leg by one of his brothers during an argument in December 2009 and was arrested when he went to hospital. He was extradited back to the UK in March 2010.

In November 2010, Mohammed Saleh Ali and Omar Hussain were found guilty of murder and sentenced to serve at least twenty-two and twenty-one years respectively.

In December 2013, Dana Amin was found guilty and jailed for eight years for helping to dispose of Banaz's body. Amin challenged both his conviction and sentence; the appeal was dismissed in September 2014.

After the case, DCI Caroline Goode was awarded the Queen's Police Medal for her work leading the investigation. Caroline was a brilliant detective for whom I had the utmost respect.

A year after the Banaz Mahmod case I was plunged into another massive forensic search that eventually led to the conviction of one of the country's most notorious serial killers, Peter Tobin, who died in October 2022.

The investigation had begun in Glasgow, Scotland, when holidaying Polish student Angelika Kluk, who had been staying at a local church where she lived rent-free in exchange for doing cleaning work, went missing in September 2006. Strathclyde Police declared her a missing person. Her sister, who worked in Glasgow, made a public appeal for information about her whereabouts. Angelika's possessions, including money, laptop, tickets home and passport, had been left in her room and her phone was missing and had been switched off.

Detectives discovered that on the day she disappeared she had helped a church handyman, Pat McLaughlin, paint a garage. McLaughlin was an alias; the man's real name was Peter Tobin, a three-times-married serial abuser, crook and psychopath who moved to Bathgate, West Lothian, in 1990. In May 1991 he moved to Margate in Kent and in 1993 to Havant in Hampshire, where in August that year he assaulted and raped two fourteen-year-old girls at knifepoint, stabbing one and leaving them for dead after turning on the gas taps in the property and fleeing. They both survived the ordeal.

Tobin went into hiding but was caught and jailed for the crime. When he was released in 2004, Tobin moved to Paisley, Scotland. A year later he was accused of attacking another woman and disappeared. He then surfaced under his McLaughlin alias and found work at the church, where he was initially questioned in the Kluk case by police who didn't know his identity or history at first. He disappeared again, and when police searched the church thoroughly, they found Angelika's body hidden under the floor, wrapped in tarpaulin. Her hands were bound and she'd been stabbed repeatedly in the body and head. A manhunt was launched and Tobin was arrested a few days later in a London hospital wearing a T-shirt with DNA on it that linked him to the murder.

He was tried in March 2007, found guilty of raping and murdering Angelika, and sentenced to a minimum of twenty-one years. The judge described him as inhuman.

After his arrest police began to piece together Tobin's movements in an effort to link him to other possible crimes, realizing they may have a serial killer on their hands. The investigation was codenamed Operation Anagram.

In 2006 a cold case review into the disappearance of a school-girl named Vicky Hamilton was set up by Lothian and Borders Police called Operation Mahogany. When it was realized that Tobin was living in Bathgate at the same time that Vicky went missing, the investigations merged and I was called in by Mark Harrison to search at locations linked to the killer. We were looking for bodies and evidence.

Vicky was fifteen when she was last seen in February 1991 as she waited for a bus near Falkirk. She had never been found. Tobin had left Bathgate for Margate a few weeks later. Police had already discovered some DNA evidence linking a purse that Vicky owned to Tobin's son. We didn't know much about Peter Tobin at that point. Our brief was to go in and search the house and gardens with the intention of hopefully finding some evidence that Vicky Hamilton had been in there and perhaps still was.

In June 2007 we travelled to Scotland and to Tobin's unremarkable semi-detached former house in Robertson Avenue. Our remit was to plan and conduct a complete forensic search of the house, gardens and outhouses. It was a huge job for us. I was briefed beforehand on the case, and as the senior officer explained how the investigation was unfolding my blood ran cold. All the evidence suggested that we were dealing with a serial killer. I quickly put aside any feelings of revulsion. I needed a clear head. It was like a switch in my brain that turned me to professional job mode. I was there to complete a task and I would do it as thoroughly as possible.

The road was closed off and barriers were guarded by police. We pulled up in our vehicles, which were loaded with sterile, individually bagged work equipment – spades, shovels, drills, picks. The house was a normal family home and was neat and tidy. Tobin hadn't lived there for years, and a family now owned it. They had moved out that morning. We wore full forensic suits and set about thoroughly and carefully searching every inch of the house, loft and garden. The whole area was divided up into a grid so we knew where to search and could log any objects or material taken for further tests. There were soil bins for the forensic archaeologists who were with us on the job.

As we worked, a neighbour told us that he remembered Tobin had dug a large hole in the garden. At the time the neighbour had asked Tobin if he was digging to Australia. Tobin replied that he was digging a sandpit. One morning the neighbour noticed that the hole had been filled in and asked Tobin what had happened to the sandpit. Tobin said he had been told to fill it in by Social Services.

On the top of where that hole had been there was now a rockery, which immediately caught my attention as a likely target because it was an unusual feature.

There were two archaeologists; the lead was Dr Jennifer Miller. They began to dismantle the rockery. When it had been cleared the human remains dog, also known as a cadaver dog, was called in. These are specially trained canines that can recognize and locate the smell of decomposing flesh. The target ground was probed with thin poles and left to vent, then the dog started to sniff the area. Within a few seconds the dog pointed out one of the holes, nose to the ground, wagging its tail.

A second dog was brought in half an hour later to sniff some fresh holes and it also indicated that there was, or had been, a scent in the ground. I ran the radar over the area the dogs had identified and got a reading that showed a large area of disturbance that went down over two metres, confirming that this was the hole Tobin had dug and told his neighbour was a sandpit.

The archaeologists started digging. They found the contours of the original hole and expertly followed the cuts in the soil made all those years ago, placing the excavated earth in the sterile soil bins for forensic examination as they went. Eventually we had a hole exactly as it would have been dug by Tobin.

When the excavation was complete there was no sign of Vicky Hamilton, but all the evidence suggested it was the site where she had been deposited for several months after Tobin killed her in 1991, before he moved her remains when he left town.

We carried on running the GPR over the garden and found more disturbed ground, which was investigated but turned out to be an area where new pipes had been laid. I also found a void under the floor in the kitchen that proved to be subsidence. There was quiet on the job – there were no radios, no banter like we had on the protest jobs. I was concentrating on the radar screen and I didn't want distractions. The team were there in their white forensic suits, concentrating fully on the jobs they had been tasked with. The atmosphere was intense and serious.

After I finished in the garden I went inside to start searching the kitchen and while I was there two members of my team, Aidan and Robin, were doing a fingertip search of the loft after removing all of the insulation, which was taken away in forensic bags. There they found several old belongings from Tobin that had stayed there for sixteen years, one of which proved to be hugely significant.

I heard a shout from upstairs and went running up. In a corner of the bare loft they had discovered a knife. It had either fallen or been placed between an end joist next to a supporting wall, in a gap about six centimetres wide and twenty centimetres deep. We carefully placed it in a forensic tube and it was taken away for tests.

We were on the site for a week. We searched the guttering around the property, we put a camera up the chimney and carefully searched every inch of the property. During that time the results of the tests on the knife came back. Traces of Vicky's DNA were found on it.

It was the primary piece of evidence that linked Tobin to the disappearance and murder of Vicky. To have played such a crucial role in the case was credit to the hard work and diligence of the team.

We had been staying away together as a team and on the final night we allowed ourselves to have a few beers in the hotel bar. While it was not an occasion for celebration, because that would have been inappropriate, it was a chance to recognize that we'd done a good job. Throughout all the criminal cases we always understood that there were human victims and lives ruined. Our work in this case would not bring the victims back, but it would help ensure Tobin would be convicted of murder, and that was the best we could ask for.

Later we were co-opted again to search another of Tobin's previous addresses in Southsea, near Portsmouth, where we cut into the floorboards and searched underneath. We also used the GPR in the cellar. We knew by then that Tobin was cunning. He liked to bury, that was his MO. And while we were there another former Tobin house in Margate, Kent, was being searched by police and a highly experienced forensic archaeologist, Lucy Sibun. It was the same story, neighbours said they'd seen him dig a big hole in the garden but then fill it in, claiming Social Services told him to do so. It was obvious to me that the bodies would be in one of Tobin's former homes. All the evidence suggested that he was a killer who liked to keep his victims near. That would have presented him with problems each time he moved as he would have had to dig up the bodies and take them with him.

It was at that property that the remains of Vicky Hamilton were found, along with those of another missing girl, Dinah McNicol. Vicky had been wrapped in plastic sheeting and still had varnish on her nails.

Dinah disappeared in 1991 when she was eighteen after attending a festival in Hampshire with a male friend. They were hitchhiking home and had the misfortune of accepting a lift from

Tobin. He dropped her friend off at junction eight of the M25 and then murdered her.

Tobin was convicted of Vicky's murder in December 2008 and of Dinah's murder a year later.

He was linked to a further fourteen unsolved murders and was suspected of many more. It has also been speculated that he was the unidentified killer Bible John, believed to have murdered three young women in Glasgow in 1968 and 1969.

I have no doubt there are many more victims of Peter Tobin buried in hidden places. He was a dangerous predator who caused misery for so many people. It makes me proud to think that me and my team had a big part in bringing him to justice.

Chapter 15

Perhaps more than any other investigation I'd worked on up to that point, the Tobin case cemented my reputation as someone who understood crime scene searches and got results.

As a team we had all grown and developed. We were arguably the most experienced private specialist forensic search organization in the country and were respected and valued by the authorities. I wasn't one for self-congratulation, so I flew under the radar, away from the media spotlight, but I did recognize that building such an outfit was quite an achievement. We used to be known as the men in black, but by the end of the noughties we were the men in white, dressed in forensic suits, working on crime scenes with some of the most senior detectives in the country.

It was the way I liked it. I wasn't in it for personal glory or for column inches. What I did for a living was in many ways a grim job. Parts of it were sickening. But knowing I had the ability to bring closure for families, and justice for victims, allowed me to tell myself that the sacrifices and the bad dreams were worth it.

And then of course there was the other side of SGI; the rescues and the protest jobs and the corporate work that, I cannot lie, provided me with a comfortable living, well beyond what I'd ever dreamed of. The work never stopped and as the years went by and we got busier, life became a frenetic rollercoaster travelling between protest camps, dive sites and crime scenes. I immersed myself in it all.

I was away for weeks at a time, and I missed my daughters, but often the absence was a mixed blessing, because at home my marriage wasn't going so well. Mandy and I were drifting apart.

I couldn't control my home life, but I could control my work, and I could dive into it and let it consume me until finding what I was looking for became the only thing that mattered.

And I had a duty to my team. I had people to support and wages to pay. I couldn't turn work down and I loved what I was doing anyway.

So, when Mervyn Edwards from my days at the Newbury bypass protest called to put me in touch with someone who was having big problems in a little coastal town in Ireland and needed a long-term solution (it turned out to be nine weeks), how could I refuse?

The problem was a clash between protesters and a partnership of Shell, Statoil Exploration Ltd and Vermilion Energy. The energy companies had formed a joint venture called the Corrib gas project, which entailed the extraction of a natural gas deposit off the north-west coast of Ireland and the construction of a pipeline to bring it ashore at Glengad, in County Mayo, and a gas processing plant. By this time governments around the world were acknowledging that gas was a better fossil fuel to use than oil and coal as it released fewer emissions. While not perfect by any means, gas was understood to be a better gateway fuel to use for energy while more sustainable infrastructure was developed.

A protest camp had been established and a range of activists had come together under an umbrella organization called Shell to Sea. The battle had been going on for several years; feelings were running high, and in 2005 five protesters had been jailed in connection with the protests. My contact at Shell explained that the company had hired a massive pipe-laying ship, *The Solitaire*, that the activists were attempting to stop at sea.

'We've got our backs against the wall,' he said. '*The Solitaire* is costing us €1,000,000 a day. It is the biggest pipe-laying ship in the world.'

Our operation and costs, while nowhere near that level, would be considerable, and it ended up being one of the biggest operations SGI had embarked on. We took sixteen people out there and while we had two boats at the time that were suitable for some of the work, we needed a bigger one (to paraphrase from *Jaws*).

I bought an ex-fisheries patrol vessel called *Nemesis*, which Chris and I refitted and painted to get it back to coded specification for the job. I also needed two extra Land Rovers with winches to tow and launch the RIBs.

I assembled a team, booked accommodation and we headed to Ireland. It was a proper military-style operation. I went out ahead to a place called Killybegs, further round the coast, where we based ourselves for a week to get settled in, to gather intel and to meet the crew of *The Solitaire*, which was sitting moored up off the coast.

Before we got involved in engaging with the protesters, we did some reconnaissance of the protest camp and watched how they operated. There were around fifty activists camped right by the shoreline. In any group there were always some who were more active and 'enthusiastic' than others. We identified them and we identified the ringleaders and watched them over a period of days. Every morning – weather depending – they got into their dinghies and paddled out to where the ships and barges were working, cutting the trench for the pipe. They then endeavoured to cause as much obstruction as possible to stop work. Sometimes they jumped onto the buckets of the marine cranes. It was a death-defying way to make a point but it did have the desired result of stopping the digging.

Everything depended on conditions at sea. The barges and dredgers could only work in a metre of swell, so during bad weather everything stopped. This made it imperative from a financial perspective for the work to go ahead when conditions were right. But calm days also allowed the protesters to get out to sea and jump in the water under the cranes. Tidal charts and weather predictions became important indicators of whether we'd be involved in action on any particular day.

When the 'fun' did eventually start it got going between 9 and 10 a.m. after we saw the activists putting on their wetsuits and flotation devices. They took to the water, and so did we. What ensued was a game of cat-and-mouse during which the protesters, of which there was a flotilla, paddled as fast as they could out to the work zone and started launching themselves off their boats into the water like lemmings. We carefully sailed around them and fished them out. Some just floated, like human obstructions, and were compliant when we pulled them out, others were active and got aggressive. The Garda was there with us too and there were several arrests. Often it was chaos.

One day, a particularly aggressive man managed to get hold of a police officer and drag him into the water, where he attempted to hold his head under. Me and one of my divers, Rick, saw what was going on and jumped in on top of the attacker, who let go of the officer and started lashing out at us. We managed to restrain him and helped to get him onto the police boat where he continued swinging punches before he was handcuffed and arrested. He filed a complaint and accused the Garda of using excessive force, so we provided witness testimony to dispute his claims.

This pattern went on for weeks. On clear days we took to the water to try to keep the protesters out of the path of the work that was carrying on. It was hard work and although the project was making progress, it looked like it would be a long-term commitment, until fate intervened.

One day a storm blew in while *The Solitaire* was working. On the back of the deck it had a large, crane-like machine called a stinger. This was used to lower sections of pipe down to the trench that had been dug in the seabed. While the stinger was on the bottom during the storm the boat moved astern, and the stinger snapped off. I heard about the incident and went into the project office the next morning where I was told by an exasperated official the exercise was over, the boat needed repairs and wouldn't be back out until the following year. In the end, with a little help from the weather, the protesters got their way.

As in so many of the protests I'd been involved in, however, the victory was temporary, and gas started to flow in 2015. But the aims of the protest movement were always more nuanced. They knew they were not going to stop whatever project it was they were protesting, be it a road or a runway or a gas pipeline. The point was to shine attention on their cause, raise awareness and disrupt. And to that end they succeeded in Ireland, where they cost the partnership many delays and millions of pounds.

The causes changed and the scale of the protests differed, but the methods were largely the same: occupy a site, invite media attention, then dig your heels in until the legal process conspires to remove you and the police, bailiffs or SGI arrive to escort you away.

While we'd largely moved away from the men in black image by the end of the 2010s, we did get involved in another water-borne operation that resembled a commando raid.

It was 2009 and a group of squatters had occupied a small island on the River Thames near Kingston called Raven's Ait Island. Under the vague banner of community activism, the group moved onto the island after its previous occupants – a wedding venue company – went bust. They argued it was common land and intended for public use, and said they were reclaiming it for the community in order to establish a self-sufficient, eco-friendly centre for community and environmental groups. The local council disagreed with their core assertion and went to court to evict them.

When the eviction notice was issued we were called to safely escort them off the island. As many of my team were ex-military and had served as Marines and Paratroopers, we looked forward to carrying out an early hour stealth raid. To add to the drama, a mist hung over the river on the night we went in.

We had plans of the buildings on the island and photography of the camp. We'd also observed it from both sides of the river so we were familiar with the layout. And we had intel that on the night we were going in, the occupiers would have been smoking cannabis and so their reactions would be slowed.

We set two boats in the water a mile or so upstream, they had quiet four-stroke engines and we chugged down the river to within a couple of hundred metres of the island, then turned the engines off and paddled our boats towards the island in the fog. Through the mist we could see a fire in the open area to the rear of the island. The squatters were sitting around a campfire. One of them was playing 'Stairway to Heaven' on a guitar, which was handy for us as it shielded any noise we might make.

We floated up to the steps that led ashore and, keeping low, crept onto the island and got as close as we could. When the song finished, I stepped forward out of the mist.

'Evening everyone,' I said.

Several of them let out a startled scream and didn't know what to do.

'You know why we're here,' I explained. 'The game's up. It's time to pack up and leave.'

A few started to argue but they soon calmed down and one of them decided to try to make a last stand. He ran to a marquee that was part of the former wedding set-up and jumped on top of it, where he straddled the apex and invited us to get him.

'Let's just leave him up there for a minute. We'll deal with him in a while,' I said as we focused on the others and made sure they got themselves together and vacated the island.

When we turned our attention back to the man on the marquee, I realized the structure was held up with poles that fitted into each other.

I laughed.

'We don't need to climb up,' I said, 'we just need to take the bottom poles out.'

And that's what we did. We each took a leg and removed the poles from the bottom until the canopy was at ground level and the squatter had no choice but to stand up and walk off sheepishly. The site was cleared easily. The protesters grabbed their belongings and we escorted them off the island. Raven's Ait had been liberated, its occupiers banished!

It was increasingly rare to get away on holiday but when I did I used the downtime to think of ways that I could improve the services that SGI provided. And it was while I was on holiday in 2009 that I watched a helicopter fly overhead and thought how convenient it would be to have a pilot's licence and access to a chopper. The amount of time spent in cars and vans going to and from jobs all over the country was draining.

I knew that helicopters were difficult to fly but I liked a challenge, so I looked for helicopter flying schools near Reigate, where I was living at the time, and found London Helicopters, based at Redhill aerodrome in Surrey.

The more I thought about it that night, and the more Rioja I drank, the more I realized how cool it would be for SGI to have our own helicopter. I could get to jobs quicker, and it would also allow me to take aerial photographs of crime scenes.

The following morning I called the school and booked a trial lesson and ten hours of flight training.

A few days after returning from holiday I drove to the aerodrome and met my instructor, Rob Garstang, who invited me into the lecture room. He seemed very dour.

'Why have you come for a trial lesson today?' he asked.

'I want to get my pilot's licence and buy my own helicopter,' I replied.

'You and everyone else,' he sniffed dismissively.

We walked outside to two small Robinson R22 two-seater helicopters on the apron. The R22 is a typical flight-school chopper that was invented by Frank Robinson, who first built one in his garage. In the early years of production there were many crashes of these machines, mainly down to pilot error, and today they are the biggest-selling helicopter in the world.

Rob pointed out the various parts and then gave me a safety brief.

'Right, let's go,' he said after the intro.

Inside the cockpit I looked at all the switches, dials and instruments and realized it was going to take some commitment. Rob called up the tower, reeled off a load of pilot talk, started the engine and wound it up to full speed. I looked out of the glass bubble in front of me. It was like being in a fishbowl, and as we left the ground the views were amazing.

We climbed to 2,000 feet and Rob said confidently, 'I'm going to show you how to deal with an engine failure in a helicopter, it's called an autorotation.'

I've come to fly, not die, I thought, but I had no time to remonstrate because a second later Rob closed the throttle and sent us towards the ground at 1,500 feet per minute. We were like a big metal sycamore seed heading towards the ground at a rate of knots. One mistake and we were both dead. As we came towards some open fields, he skilfully put the helicopter into a hover and then took off again. We did some steep turns and he explained the instruments and how if you don't lower the collective quickly enough the blades will stall and you crash. It almost appeared as though he was trying to put me off and I wondered what I had got myself into.

As we headed in to land, he did another simulated engine failure and sent us hurtling towards the ground again before we hovered back to land.

In the debrief room he asked if I still wanted to be a helicopter pilot. I said yes and told him that I'd booked ten hours and was flying with him again the next day.

To gain a helicopter pilot's licence I needed to pass nine exams, fly a minimum of forty-five hours, take a stringent pilot's medical and take a final flight test. The whole process cost in the region of £20,000.

As my hours in the helicopter built, I got to know Rob's dry sense of humour and we started to get on well. He advised me to read a book called *Fatal Traps for Helicopter Pilots*, which is about deadly helicopter crashes, so I could familiarize myself with some of the common and not-so-common mistakes.

My exams flew by, and I passed each one as my hours built. With each passing milestone I got more confident. I was ecstatic to complete my first solo flight, a circuit around the airfield, and get my solo certificate, which I took round to show to my parents like an excited schoolboy. While they were full of praise, when I went home and told Mandy, she showed no interest.

That spurred me on and the following day I flew five circuits solo. Over the next few days I completed ten hours of circuits before the next challenge, which was ten hours of cross-country flying away from the airfield on my own. Before I could complete this I had to show the instructor that I could talk confidently on the radio, fly the helicopter safely, land it if the engine failed, navigate without GPS and fly into controlled airspace. My first cross-country flight to Guildford and back went without a hitch and over the next few weeks I completed that element of the training.

Next was the 'land away' cross-country, where I had to fly to and land at two other airports. I was given Rochester and Lydd in Kent. I planned my route, worked out the winds and marked up the maps, checking the weather and the Notices to Air Missions to ensure the Red Arrows were not going to buzz me en route.

I completed the flight without fault, stopped for a cup of tea and a cake in Rochester and negotiated some congested airspace over Kent. It was a big achievement for me and over the following weeks I brushed up my skills for my final flight test.

The night before I went to bed early and slept through. The next morning, after a good breakfast, I drove to Redhill Aerodrome where I met Will, the examiner, who was a highly respected pilot with thousands of hours under his belt. In the classroom he questioned me on all aspects of flying and asked me to prepare the route and do all the appropriate planning for the test. Outside he walked me around the helicopter, asking me how everything worked, and watched me carry out my pre-flight checks.

The test took two hours and involved cross-country flying and a range of simulated engine failures. In the cross-country section I was tested on a navigation instrument called a very high frequency omni-directional range, which pilots used before GPS to fly from beacon to beacon. Back at the airfield there were emergency drills and false landings. It was an intense 120 minutes during which I needed optimal concentration and a calm head to deal with the next challenge that was thrown at me.

I was concentrating so hard it seemed to go in a flash and when Will asked me to take the helicopter back to the hangar and land it, I was surprised the test was over. I shut the engine off and he put his hand out and said: 'Congratulations, Peter, you've got your wings.'

I struggled to hold back my emotions. I was so pleased. It was a huge achievement for me and represented a new phase of freedom. My parents had come along for support and had waited patiently. They were eager to hear my news and were delighted when I walked towards them beaming.

I went home and told Mandy.

'Well done,' she said.

Soon after I bought SGI's helicopter. I had my own version of Tracy Island, with boats, cars, 4x4s, a command centre and motorcycles.

We became known for using cutting-edge tech. Since those early days when I brought SSS to the UK and started using it for underwater searches I had followed developments in technology, always on the hunt for the next big thing. All kinds of kit was being developed, particularly in the robotics and drone industries, some of which seemed too good to be true. One day I got a call from a man who told me he'd invented a device that could look through walls. I was sceptical but I invited him to come to the offices and give me a demo. If anything, it would give us a laugh. He turned up with what looked like a gun with a couple of aerials on the top. He claimed it could detect a heartbeat through rubble. It felt lightweight, like a child's toy. He said he had various government departments interested in buying it. I was not convinced and politely declined, but years later I found out he'd been arrested. The government had bought this thing and put it through extensive testing. It never worked. It was a scam.

The helicopter allowed us more freedom to get between jobs and my parents, who loved hearing stories about the things I was getting up to, were very impressed when I took them for a maiden flight.

Other jobs came and went. There was a protest at West Burton power station where a group of mainly female activists cut through the security fence like commandos and set up a tent on top of one of the chimneys. I jumped in the helicopter and flew straight to the scene.

There was also a protest against the new Bexhill relief road in Sussex over the Christmas holidays, in which hard-line protesters in trees wearing black balaclavas fought with unprepared bailiffs in the snow. My team went in, scaled the trees and managed to get them all out safely, and also cleared a tunnel under a second camp. That protest got a lot of media attention because one of the most vocal activists was the daughter of Chrissie Hynde, the lead singer from The Pretenders.

Our facilities and experience meant that in 2012 we were awarded a groundbreaking contract to provide twenty-four-hour support to Surrey Fire and Rescue Service for operations where we had unique expertise, particularly in swift water, rope and confined-space rescues. The decision, passed by the county council's cabinet, was the first time a private firm had been contracted to provide emergency response in the UK. Under the agreed terms we were authorized to drive on blue lights, and we did a week's training to enable us to do so. We were allowed to drive twenty miles an hour above the legal limit of the road we were on. At the end of that course, I had to do a final twenty-five-kilometre drive on blue lights to get my ticket.

It was high-octane stuff, but nothing impressed Mum as much as a job I was involved in soon after I passed my helicopter licence.

'I'm going to see the Pope,' I told her one day.

Mum was Irish and a Catholic.

I'd been approached by the serjeant at arms at the Parliamentary Estate to take part in a highly sensitive project. I was well known and trusted by the authorities. It was 2010 and Pope Benedict XVI was making a state visit to the UK. It was a high-profile event and the authorities did not want the embarrassment of any protesters disrupting the Pope's scheduled visit to the Houses of Parliament.

To monitor for protesters and nip any problems in the bud, me and a team stationed ourselves on the roof of the Palace of Westminster for a couple of days and nights to keep watch. During that time, we were part of the wider security operation that included snipers and anti-terrorist police, who were all low profile and in the background. We had full access passes to the Elizabeth Tower, which houses Big Ben, and all the other iconic buildings, and we kept patrol to make sure there were no nasty surprises when His Holiness arrived.

On the day of the visit everything ran like clockwork. There were armed diplomatic protection officers everywhere on motorcycles. Our team was stationed up on the roof and I could see the Popemobile as it drove into the private courtyard of the Houses of Parliament. I couldn't resist getting a picture for Mum. It was amazing to see such a famous man who commands so much power as the head of the Catholic Church. I felt privileged to be part of the day.

Chapter 16

Once when I was explaining to someone at a party that as part of my work I helped find lost remains so families could get closure, they said: 'So you work in God's lost property office.'

I thought that was a poetic way to describe the most sensitive and often the most difficult aspect of my job. On searches my forensic, investigative mind is always focused on the scientific and the factual. Searching is about using the right tools, looking for the clues and unlocking the puzzle. That's the objective side. But I'm also a human, with a family, and I empathize with people – I'm not immune to the pathos of some of the work I do. While I can dial down the subjective side of my mind to allow me to concentrate, the human element is never far away. It is not just about retrieving a body from a river or finding buried remains. Behind every item of 'lost property' I manage to recover, there are human stories of love and loss.

In 2010, two very different searches showed just how important and emotive my work can be. In June, I was called by West Mercia Police to look at Operation Rainbow, which was an eighteen-month-old missing person case that police were about to close and put away without conclusion.

The subject of the search was Damian Tudge. The forty-one-year-old father-of-two went missing on 13 November 2008 after a night out drinking heavily with his wife. When they returned to their Kidderminster home they got into an argument and Damian, a welder, hit his wife. She later told an inquest that the attack was totally out of character and that in the twelve years they had been together, Damian had never laid a finger on her. She explained that as soon as he realized what he'd done, Damian went white with shock. There's no excuse for hitting a woman but Damian's reaction and the punishment he meted out to himself was disproportionate. He felt so bad about what he did that he left the house, got in his car, drove off and at 2.39 a.m. on 14 November he sent a message to his wife. It read:

I love you and I will always love you and the kids but after what I have just done, I am going to do the right thing.

It was the last time anyone ever heard from him. He disappeared and his blue Peugeot 205 was never found.

Cell site analysis from that last text showed that he'd been in an area near Bewdley Town Football Club, close to the River Severn, so that's where I started looking. I began by surveying the Ordnance Survey map of the area to find any mineshafts, quarries, tunnels or lakes where a car might have stayed hidden for over a year, but the only thing that stuck out was the river.

This was a puzzle because the police were adamant his car was not in the water. They based this assertion on advice from the Environment Agency, who said if Damian's car was in the river, then oil and petrol residue from it would show up on their sensitive monitoring equipment. I was also told that the area of the river had been searched several times with a special sonar device.

This immediately sent my spider senses tingling. I'd done enough secondary searches to understand that what police forces sometimes thought was an effective apparatus was usually the opposite.

'Who had the sonar that was used?' I asked.

I was told it was a voluntary search and rescue organization. Police forces regularly use voluntary organizations to carry out surface searches of rivers as their own numbers are greatly depleted, and while these people do a great job and should be commended, the equipment they use is never up to the spec that we have at our disposal.

'I can guarantee they used a fish finder,' I said.

Fish finders do what it says on the tin. They are cheap sonar units that fishermen use to find shoals of fish. They are not good for finding submerged cars or bodies.

I called the head of the search and rescue unit who confirmed that they had used a fish finder on their searches. I thanked them for their input and help. The call confirmed my assumption that the river had not been searched properly and was the most likely place to find the car, and possibly Damian.

There were no slipways into the river that he could have driven down, so it was likely he'd driven off the bank. As he had disappeared eighteen months previously, any damaged bushes or uprooted vegetation that might have provided a clue as to where he went in had long since regrown.

Luckily there was a weir and a lock downriver from where the cell site analysis suggested he'd been. A floating car, or a moving submerged car would not have been able to get through it. This gave us about a mile of river to search.

I had my dive team with me and we got kitted up, worked out

a grid, launched the inflatable boat with the SSS and began the search. Within fifteen minutes I located not one but two cars on the riverbed.

One of our divers, Aidan, got in the water and swam down to the first. Visibility was poor and he felt around the vehicle to locate the number plate, which he removed and brought to the surface. It was not the plate for Damian's car and when the police who were with us checked it, they discovered it was a car that had been stolen years earlier.

The other car we'd found was further up the river and close to the road on a corner. We set up on the bank and used the surface demand diving equipment – this is equipment that provides breathing air to the diver through an umbilical line from the surface, rather than through a tank on their back. The idea was to swim down and take a camera to film the plate and then identify the vehicle from that. I was the first diver in, rolling off the side of the boat. Chris passed me the underwater camera on a cable that transmitted pictures to a screen on the surface. I descended slowly down the shot line but could see nothing. The water was filthy. I swam around in the gloom until I touched the roof of the car, then around the car to feel the windows. They were all closed and intact, which suggested that if Damian was in the car when it went in, which was highly likely, he was still in it. I tried to look through the windows but it was impossible to see anything clearly inside. There was no question of trying to open the doors or smash the windows, for two reasons. Firstly, the car was a potential crime scene. Although the evidence suggested Damian had taken his own life, we could not be 100 per cent sure until we checked the vehicle and so I didn't want to disturb the scene any more than was necessary. The second reason was that Damian was probably inside; I didn't know what condition he'd be in but I'd seen enough bodies in water to know that it wouldn't be good. And I didn't want to open the doors or windows only for bits of him to float away.

Having completed the initial reconnaissance, I swam to the surface to brief my team. My colleague Rick took over and went down to get the number plate, which he brought to the surface. It too was covered in a layer of algal grime and silt which was wiped away to reveal the registration that we knew belonged to Damian's car. We'd found him.

Once we knew that we had our target, we notified the police and the next stage of the operation then became recovery, for which a crane was brought in and the road was closed off. This was not something that the public needed to see.

In normal circumstances a car would be lifted horizontally with chains passed through the windows and under the roof, but we couldn't open the windows as we had to preserve evidence. Rick and Aidan dived and secured the chains around the axles and wheels.

Very slowly the car was lifted to the surface and emerged from the water, back end first. As it rose free, the water inside drained away. Mercifully it was still impossible to see inside. Out in the daylight no blue paintwork was visible. The car was mucky brown. Few understand how an object changes appearance underwater in a matter of months and a visual search may well have dismissed this vehicle because now it was brown if viewed on an underwater camera from the surface.

The case also reinforced one of my work mantras, that you should always question what others tell you. Had I accepted the Environment Agency's declaration that if there was a vehicle in the water the oil and petrol would have been picked up by their monitors, I would also have discounted the likelihood that the car was in the river.

Once clear of the water the vehicle was lifted over the bank and over the road. At this stage I was concerned that if Damian was in the car, he was now likely to be resting against the windscreen. I didn't want it to give way and for him to spill onto the road. We carefully worked with the crane driver to get the Peugeot down onto the tarmac on its roof and then rolled it over. I could already tell from the smell that there was something decomposed inside. Up close we could see properly inside for the first time and identified Damian's body in the back. It was well preserved, but he'd floated around in the vehicle. The next task was to open the doors, recover the body for the coroner and check and record the controls inside the vehicle for evidence in case a crime had been committed.

The doors were fused shut so we cut the locks off. Inside I looked to see if there were any blocks on the pedals that would denote murder. I checked where all the controls were because if the ignition was off or the car wasn't in gear, this could suggest foul play. But the ignition was on and the car was in fourth gear, which suggested he'd driven at speed before crashing through the bank and into the water. The demist was also turned up full. I imagined him sitting there on that cold November morning, in the dark, confused and mortified by what he'd done before he sent that last text and ended his life.

We very delicately removed his body and placed it in a body bag quickly, as the flies were already swarming. CSI found his mobile phone in his pocket.

It was a sad end to a life cut short and it was not a pleasant task. It was never pleasant recovering bodies, but I gained some comfort in the knowledge that we were able to bring some closure to the family.

Several months after recovering Damian Tudge's body and solving the sad mystery of his disappearance I was called again by the national search adviser to help with another missing case. This one was confusing.

The missing person was Hieronim Jachimowicz, known as Henry. The eighty-seven-year-old Pole had not been seen since March 2005. Up to that point he lived in Borehamwood, Hertfordshire, with his son, Michael, a pensioner in his late sixties who was his father's sole carer.

When Henry disappeared, Michael variously told neighbours and social workers that his dad was alive and had gone to live in Lithuania, Poland or Bristol. However, Michael had added his father's details to the headstone at the family burial plot in Gunnersbury Cemetery, West London, where his mother was buried.

One neighbour was suspicious at the inconsistency of Michael's explanations and the fact that Henry had gone without mentioning a word of his plans. She repeatedly went to the local police. She didn't believe her neighbour and suspected he had killed his father and buried him in the well-kept garden. She badgered the police for five years and eventually got a solicitor to write to the chief constable, who then tasked the Specialist Crime Directorate to start looking at the case properly. When they did, they began to see inconsistencies that set off alarm bells. A delve into Michael's bank accounts showed that he had accumulated many thousands of pounds since his father's disappearance by continuing to draw the old man's pension.

This was enough to set Operation Provost into action. The plan was to surprise Michael one morning and take him away for questioning, allowing me and my team to search the property.

To this end I arrived along with my search team on a chilly November morning in a quiet residential area of Borehamwood where we waited in our van, out of sight, until Michael had been taken away. I later heard that he wouldn't open the door to the police, so they popped it and arrested him. I also heard that he was put up in a hotel while we continued our work as he wasn't seen as a threat to anyone. Apparently, he loved his dad, he loved his animals and loved his house and garden.

When we arrived at the scene it appeared to be a normal well-kept home, with an immaculate garden full of fruit trees and plants. Michael and his father were obviously big gardeners. There were empty beehives, which at one point had been tended lovingly by the old man, before he became too frail to do so.

I did a cursory search of the house and garage in which I immediately noticed a couple of hand grenades. Normally I would have been surprised but the background details I'd been given on Henry said he was a former Second World War soldier, and I realized straight away that the grenades were deactivated and not a problem. No doubt they were just an old soldier's souvenir of active service.

We went into the garden with the team. At the scene we were working with Mandy Chapman, who oversaw the Met's human remains dogs, and forensic archaeologist Dr Karl Harrison and his team.

I scanned the area and tried to get into the mind of someone who might be planning on burying a body. This was a technique I'd learned through experience.

'Where would I put him if I were him?' I asked myself. The most logical place would be in the least-overlooked part of the garden. It's not something you'd endeavour to do in the middle of the lawn.

The row of beehives were sheltered behind bushes, and the space underneath them would have provided room for several bodies. That was an obvious area to scan. I also ran the radar over the patio, just in case. The SIO was convinced that's where we'd find the body but I doubted it. Burying a body under the patio is another big undertaking, and from a psychological stand-point, people generally prefer to dispose of bodies somewhere remote, not right outside the back door. The radar confirmed the patio was clear. I suspected the SIO may have been watching too much *Brookside*.

Having marked out where I thought the most probable areas were, I did a scan of the entire garden and found a possible target area. The radar showed a disturbance several feet down. The ground was probed and vented and one of the dogs was brought in. It indicated there was something there straight away. Karl then started to dig down and saw cut marks in the soil where the ground had been previously dug. He excavated further and found a layer of potting compost, which was unusual. Then he uncovered something solid. At that point I'd only been at the scene for less than an hour and thought to myself, 'I'm good, I've already found what we are looking for.'

As the digging continued, we found wooden boards. I knew it was traditional in some Slavic cultures to create a structure over a casket using concrete blocks at the top and bottom and planks on top to prevent it getting crushed by the soil.

'This is a classic Polish burial,' said Karl.

The anticipation was palpable as the plank was moved aside to reveal a casket. There was something not quite right, however. The box was only a few feet long.

'Either the man was a dwarf, or he's been dismembered,' I said.

Great care had been taken with the burial. The casket looked handmade and had been tied and sealed. Karl prised open the top. Inside, wrapped in plastic, were the remains of a dog.

Gallows humour is one of the most important ways people who deal with awful situations in their jobs process things and get through them without going crazy. You find it in the police force, the ambulance service, the fire brigade, the military. Sometimes all you can do is try to find humour in the situation, and I'm not ashamed to say that when the lid was lifted and 'Fido' stared up at us, there were a few guffaws of laughter.

'He must have been barking mad to go to all that effort for a dog,' someone said.

We lifted the dog coffin from the hole and checked underneath, to make sure it wasn't a decoy and then did the decent thing and put it back in and reburied it.

After that early bit of excitement, I carried on with the radar. The SIO was still adamant that the patio be dug up.

'It's several days of work,' I explained.

'I'll get my officers to help you,' he offered and called in the bods from the Tactical Support Team who came in with Kango hammers and demolished the area. I watched, with head in hands as I knew there was nothing there because the radar would have picked it up straight away. Sure enough, when the dust settled, the area that was once a patio had been thoroughly excavated and was deemed free from buried remains.

I moved on to the hives, where my initial suspicions had led me. They were sitting on patio slabs, so we moved them aside, lifted the slabs and scanned the area. The radar showed another shadow in the ground, this time too small to be a buried body – unless the father and son had kept hamsters as well as a dog.

I dug down anyway, in case the buried item was evidence, and uncovered three buried revolvers and 200 rounds of ammunition. The package had been carefully wrapped up and I could see it was old Second World War stuff. More souvenirs. To the untrained eye this would have been a highly suspicious find but I was confident that these had been buried to keep them out of the way of anyone who could have used them for harm. Whoever buried them – either Henry or his son – probably thought they were being responsible.

The items needed to be handled as evidence in case the job turned into a criminal investigation. As I was bagging them I heard a shout coming from inside the house.

'Hand grenade! Evacuate!'

Someone in the garage had found the grenade.

I got up, ran inside, and explained.

'It's okay, it's a dud, it's deactivated.'

But the SIO was adamant that safety came first. All work stopped and an Explosive Ordnance Disposal (EOD) team were called in. Their base was in Didcot, near Oxford, over sixty miles away. They arrived with sirens blaring an hour and a half later. The sergeant walked in. I told him I was ex-military, and that the bomblet was an old, deactivated pineapple grenade. He looked at it, picked it up, examined it, and confirmed that it was harmless. The bomb squad got back in their vehicle and drove back to Didcot. We got back to work.

An hour later there was another shout.

'Evacuate! We have two grenades in the top bedroom!'

Here we go again, I thought, as the site was shut down once more. It was turning out to be a long day. Once more the EOD team turned up an hour later, confirmed that the grenades were harmless and left.

We had been on site for several days and towards the end of the week it started pouring with rain, which hampered my progress with the radar. The SIO decided that with no more positive results from the radar or the dogs it was time to wrap up the operation. But I hadn't finished and explained that I couldn't go without being satisfied that I'd thoroughly checked every conceivable area.

One place I had yet to check was under the garden path, which was made with slabs. The dogs had run over it all week without registering anything, but I knew that a dog would only pick up a body if it could detect a scent. And they couldn't smell through concrete.

There was a barrel on the pathway, which we moved. I used another bit of equipment called a magnetometer, in addition to the radar, which measured magnetic strengths and detected buried artefacts.

Working together, the apparatus showed a disturbance. The side elevation scan of the area indicated a human-sized target just under a metre beneath the ground. It could have been another dog, but why was it buried under a path?

As it was raining, we put a canopy up over the area. We removed the heavy slabs and Karl started to dig into the ground. He found the cuts from the original dig and followed them down. He found the compost again, and some road scalpings.

Then he found small tree roots, which he dug under. A few centimetres under the roots he uncovered an old plastic milk container. Inside there was a note. It read:

Henry's dying wish was to be buried with all the things that made his life worth living, viz: his beloved wire fox hair terrier (three buried in this garden), his wonderful bees (six hives here) and his glorios [sic] garden of twenty fruit trees.

He held me to promise that come his final day on the good earth, that I would ensure his request be met.

To that effoc [sic] I have honoured his wish.

Found dead in bed on Sunday morning 13/3/05.

All my love and lifelong respect. God be with you!

Signed, Michael, his only son.

P.S. I write this with tears in my eyes.

It was an emotional explanation to describe what Karl unearthed next. There was a sheet of tarpaulin in which the body of a little man was wrapped. We lifted him out very carefully. He was mummified. We put him in a body bag and carried him into the garage out of the rain where a doctor was needed to formally declare him dead before he could then go off for any forensic tests or autopsy, if that was needed.

A young doctor turned up about an hour later. She had a stethoscope around her neck. 'Where is he?' she asked when she walked through the front door.

'I'll show you,' I said, leading her to the garage. 'You won't be needing the stethoscope.'

'I'll make that judgement,' she said curtly.

We arrived in front of the body bag and I crouched down and looked up at her.

'Are you ready with your stethoscope?' I asked.

I then unzipped the bag and opened it up. She recoiled slightly at the sight. She was not amused.

The job had been successful, we'd found Henry, but rather than it being a case of something sinister, it appeared to be a case of a son's love for his father. I could appreciate the emotions that Michael would have felt as his dad made him promise to bury him at home. Michael would have wanted to be with his father but would also have been frightened of the repercussions, knowing that it was illegal to bury someone without authorization.

The justice system didn't see it that way and, to be fair, Michael had done himself no favours by fraudulently claiming his dad's pension. He was charged with manslaughter and went to trial at Snaresbrook Crown Court in July 2011. The prosecution claimed he'd covered up the death in order to gain from it financially.

According to reports Michael maintained he had concealed his father's natural death on 15 March 2005 so that he could honour a pledge to bury him at home. Asked why he had not called a doctor, he said: 'Father would have been taken to the local hospital, then from the hospital he would be put in a mortuary and then to the undertakers and then to the cemetery.

'Now I have a problem facing me. I thought, "What am I going to do? Am I going to allow that to happen or am I going to honour the pledge I gave him to bury him in the garden?"'

He told jurors: 'I slowly realized I had made a promise to him and that promise was that I would bury him in the garden.'

Jachimowicz told the court he spent the entire day digging

the grave and had been to a nearby DIY store to buy 'the most expensive tarpaulin' to wrap him in.

After dark he picked up his father's body and carried him to the grave.

'Then, as gently as I could, I placed him in the tarpaulin,' he said.

The jury believed him, and he was cleared of the manslaughter charge but was sentenced to a twenty-month jail term on 2 September 2011 for burying his father unlawfully. But that wasn't the end of the story. A week later he was back in court. This time he was ordered to forfeit £36,602 for claiming his father's pension up to six years after his death and was also ordered to pay £25,546 in compensation.

I did hear later there was a happy ending of sorts. He was allowed to rebury his father in the garden, which is what father and son wanted in the first place.

Chapter 17

In an ideal world a crime scene should be a hermetically sealed capsule where everything coming in and out is controlled and monitored; where every inch is searched and where every speck of dust, skin flake or splash of bodily fluid is tested. The world is far from ideal, however, and crime scenes happen everywhere, sometimes in the most chaotic and exposed places, from back gardens to busy streets. People traipse through them and unwittingly disturb evidence.

These are the physical outside influences that investigators must contend with. And then there is also human opinion, which often skews investigations and leads people down dead ends. What gets searched and how thoroughly the search is conducted is often down to individual opinions and assumptions of those in charge of the crime scene. In my experience that is often where things can go awry. We are all fallible. No one is perfect.

The trouble being that if you get things wrong on a crime scene and evidence gets missed, criminals get away, justice is evaded, loved ones get no closure.

I learned these lessons during one early case, back in 2003, when I was asked by a SIO from the Met Police to search a house and gardens in Ealing, West London, for a freelance photographer, John Goodman, who had been missing for nine months.

He was a cancer patient and had been reported missing by a neighbour. When police knocked to check his whereabouts, his door was answered by Fabio Pereira, a Brazilian drifter who, it transpired, had squatted in the house when John was away. Pereira was discovered when John returned, who then murdered him. He then embarked on a £10,000 spending spree using John's credit cards, mobile phone and funds from his bank account.

He used the money to pay for champagne, lap dancers and nightclub visits. While investigators were able to build a case against Pereira and charge him, they couldn't find John's body.

I was called in after Pereira had been arrested and was on remand. I was tasked with searching the house; a Met search team was already there but they didn't have the advanced technology that I had. There was also an archaeologist working in the garden who told me there was no need for a GPR scan of the area as she had already surveyed it and there was no disturbed ground to check.

After intervention from the SIO it was explained that I had been brought in with my team and that the equipment was more effective than a visual search.

'Thanks for your advice, but we're going to carry on with what we've been asked to do,' I said to the archaeologist politely.

The back garden was gridded off to carry out a systematic search. I walked the radar carefully following the lines that we had pegged out and hit a target within minutes. The reading showed a big reflection that indicated a buried metal object. The archaeologist insisted that she dig it up. It was a crowbar and a hammer.

'That's embarrassing,' I thought to myself.

We continued the work. There was a small ornamental well

in the garden which we pumped out with a trash pump. It was full of crab apples that shot through the pump like a peashooter into a sieve to ensure we lost no evidence. There was nothing at the bottom and once we'd scanned the rest of the garden and were satisfied there was nothing there, we went inside where the Met search team were working. I scanned the floor of one of the rooms that showed something else metal, but it was two buried silver trays under fresh concrete. We excavated the floor to ensure the trays were not covering something more sinister.

As it was a big old house I wanted to get in under the floorboards to search with the radar as I knew that voids under floors were perfect places to hide bodies.

'It's done,' I was told by one of the police search advisers.

'Well, we need to check it anyway. I'd like to get the GPR under there and the rest of the house,' I reasoned. But I was told that SGI, our GPR and myself were not required as a thorough search had been conducted. A few years later I probably would have stuck to my guns and made a fuss but there was only so much I could do in a crime scene and if those in charge didn't want me to search a certain area, or told me that it had already been searched, I had to respect their wishes. We left the job, trusting that all the underfloor spaces had been thoroughly searched.

In total there were three searches of the property and dogs were used twice but John was not found and the case went to court without a body, enabling Pereira to insist that John was still alive.

In October 2004, Pereira was tried and found guilty of murder. A week after the verdict, before sentencing, he admitted to the killing and claimed it was an accident. He told police where the body was. John's remains were hidden under floorboards. They were badly decomposed but they enabled detectives to establish that John's hands had been tied behind his back, his shirt was pulled over his head and knotted and a surgical stocking had been used as a gag. Pereira was sentenced to a minimum of sixteen years.

It was one of those frustrating cases where opinions differed and where, if things had gone right initially and I'd been allowed to search where I wanted, the family would have been spared many months of uncertainty.

Over time, as I attended more crime scenes and got results, I was also called upon to conduct tests, particularly when cases involved confined spaces. I provided forensic analysis in several cases similar to the body in the barrel murder, where I was asked to simulate death scenes under controlled conditions to ascertain data and test theories about causes of death.

In one tragic case, coroners were trying to establish what caused the death of a baby who died in a bed next to its mother. It was what they called an overlay case. The mother was a drug user and slept with three duvets because her addiction made her permanently cold. Her child was found next to her, dead in the bed. My job was to establish if somebody could suffocate underneath three duvets. I reconstructed the scene at our offices and set up a bed with duvets of the same tog, which I got under with an oxygen monitor, a temperature gauge, and a pulse oximeter to measure my blood oxygen levels. At intervals an assistant recorded all the data from the monitors on a spreadsheet, which was then sent to a paediatric respiratory expert to be analysed further and adjusted to take into account that my rate of respiration was obviously different to that of a baby's.

The conclusion we arrived at was that it was possible to suffocate under a duvet. Over time there is a massive build-up of carbon dioxide, which displaces any oxygen and drastically reduces the oxygen levels from 20.9 per cent to dangerous levels where life becomes extinct. I gave evidence of my findings in that case to a coroner's court in Liverpool.

Not long afterwards, I was asked to assist in another tragic investigation involving a child. The poor mite didn't have a chance during her short life. Her dad was a drug dealer and her mum was a sex worker. She was neglected throughout her short life and was found dead in a divan storage drawer under a bed with a sock stuffed in her mouth to silence her.

It was one of those horrendous cases that hit particularly hard because I was the father of young girls and could not fathom how someone could do that to a child.

In that case I was called to meet detectives from the Met Police and scientists from the Forensic Science Service at their Lambeth HQ. They wanted me to reconstruct the scene to establish the likely cause of death. I was contacted because of my previous work. One of my team created an identical bed at our offices to the exact dimensions of the bed in which the little girl died. We coated our reconstruction in plastic, so it was transparent, then put it next to a radiator, as that was where the original bed had been located. We conducted the same set of tests as with the duvet suffocation, with my team member as the guinea pig in the makeshift divan, this time using twin-calibrated oxygen monitors to measure oxygen depletion over time and a temperature sensor to record the temperature changes within the confined space.

The experiment showed that it was unlikely that the girl had suffocated. The temperature rose rapidly, however, and the most likely scenario was that the victim overheated. It was a horrendous way to die. My evidence was given to the Old Bailey and the parents got five years, but in my opinion that was not enough. They were evil and deserved much longer.

We also did a similar test after a young woman was found dead, naked in the back of a car in a London park. There was no sign of a struggle, there were a couple of empty water bottles with her and she was known to have had mental health problems. A pathologist thought she may have suffocated but a post-mortem was inconclusive.

We used the same model and set up our equipment in one of our industrial units, placing the subject inside. That time, Chris was the test subject and although we tried to simulate the situation as closely as possible, we asked him to keep his clothes on as the remit was to test whether it was possible to suffocate in a car, so it didn't require him going Full Monty. He couldn't open the door and get out, though, because that would affect the oxygen and temperature levels, so he was provided with a plastic bottle, should nature call.

Like the trooper he is, Chris stayed in there for nine hours to illustrate that the oxygen levels remained largely stable, although the test did show that the temperature went up by ten degrees, even though the car was in the shade. The woman's car had been in direct sunlight. This led us to the theory that she had dehydrated and overheated, then became delirious and stripped naked before she died.

My experience in scientific tests of confined-space cases such as these eventually led to a two-year involvement with one of the most perplexing police investigations of modern times; the enduring mystery of the spy in the bag, a case with more layers than an onion that has left me deeply sceptical of the official line to this day.

It began back in August 2010 – a year when the body count of cases I was involved in stacked up considerably. It had been a few weeks since we recovered Damian Tudge from the River Severn and I'd gone away to Spain with my family to try to relax.

While I was there, I got a call from the Specialist Crime Directorate of the Met Police asking if I could have a look at a crime scene at an address in Pimlico, a region of London near to the MI6 headquarters in Vauxhall, which has famously featured in several Bond films.

'It's quite urgent,' the contact said. The story was headline news. The victim was a man named Gareth Williams. He worked for GCHQ and was seconded to the Secret Intelligence Service.

I immediately felt an urge to get back to London and get stuck into the case. It sounded like dynamite.

Gareth was thirty-one when he died. He was a maths prodigy and had been working at the top-secret government listening and signals intelligence service in Cheltenham since 2001 before he was moved to MI6 in London after receiving training for 'active operational work'. In the city, he lived in a flat close to the Secret Intelligence Service HQ. He had requested an early return to Cheltenham because it was reported that he disliked city life. His sister later described him as a 'country boy' who hated the 'rat race, flash car competitions and post-work drinking culture'. He was a keen cyclist and walker and was due to return to GCHQ in September.

He had few friends. His landlady in Cheltenham for ten years said he never had anyone in his flat there, and he was described as a 'scrupulous risk-assessor' who was as meticulous as a 'Swiss clock'.

Gareth had not turned up for work for several days and had not been in touch with anyone. Eventually, concerned colleagues reported his disappearance to police, who made a 'welfare check' at his flat on 23 August 2010. There was no answer, so they entered the property, which was described as 'extremely tidy' with no signs of disturbance. Gareth's mobile phone and two SIM cards were laid out on a table and a laptop was on the floor. An officer noted a lady's wig hanging from the corner of a kitchen chair. Everything else seemed normal, until they searched the en-suite bathroom of the main bedroom. In the bath there was a bulging red North Face holdall with the zips padlocked together. There was also a peculiar smell. One officer lifted the bag, and fluid seeped out. He realized there was a body inside. It had been there for a week and was badly decomposed.

Forensics confirmed it was Gareth and there was no evidence of a struggle and no fingerprints in or around the rim of the bath.

Within a week of all this happening, I was back in the UK and walking up the stairs to the flat where I met Metropolitan Police homicide and serious crime investigations detective Jacqueline Sebire, who oversaw the case.

In the past, the crime scenes I had attended were full of activity, with forensics officers, CSI experts, archaeologists and police investigators. But this scene was weird. Everyone had gone and there was just me and Jacqueline and two National Crime Agency (NCA) officers. All the trappings of a big forensic investigation remained, however. There were blocks on the floor that acted like stepping stones to allow personnel to walk through a crime scene without contaminating the floor. The bathroom was still covered in fingerprint dusting powder and Luminol – a compound used to identify blood and other body fluids under ultraviolet light – coated every surface. The bag containing Gareth's body had been taken away, but the unmistakable stench of death still hung in the air. No amount of crime scene chemicals could cover it.

I carefully looked around and assessed the scene. I noticed the wig, which was still there. I couldn't see any signs of a struggle and the dusting powder around the bath showed no fingerprints.

I was shown photographs of the bag with Gareth's body inside and close-ups of the Yale padlock. When Gareth had been found, the heating was turned up to full, the shower screen was closed, and the bathroom door was shut. The front door had been locked from the outside. It was all highly suspicious.

A few days later I had a meeting with the Specialist Crime Directorate in Hendon to discuss the case. I asked questions.

Was there any CCTV from the building? Who else lived there? What kind of work was Gareth involved in?

At the meeting various scenarios were discussed. Was it a hit carried out by someone with good forensic awareness? Was it an elaborate suicide? Was it stage-managed to look like a sex game gone wrong? But the scenario that in time gathered the most momentum was that Gareth was on his own and it was a kinky sex game that went wrong. In other words, to get his kicks he'd locked himself in the bag in the bath and suffocated. To be honest, it seemed implausible to me. I know people are into auto-asphyxiation and bondage, but the technicalities of what was being suggested seemed impossible. But I endeavoured to keep an open mind.

It was explained that there was £15,000 of high-end women's clothing found in the flat. The inference being that Gareth had a complicated private life.

The best way to test the theory that he'd died by his own hand was to see if it could be done, if a man of Gareth's five-foot-eight-inch, nine-stone stature could padlock himself into a bag in a bath without leaving marks.

I began running reconstructions with me playing the part of Gareth, as I was of similar frame. We also staged reconstructions at work where I tried to get into the same make and model of bag. I found it impossible. I tried time and time again. I was then zipped in by the team, with my knees up to my chest and my head bent down at a neck-straining angle. Once in, I recorded controlled environment tests on oxygen levels and temperature. I had a small knife around my neck in case I needed to cut myself free and there was a medic on hand in case I passed out.

It was unbearable in the bag, hot and claustrophobic. I couldn't imagine how awful it would have been with the heating on. It was so cramped; it was hard to read the oxygen metre. Through natural respiration I managed to deplete the oxygen within the tiny space to dangerous levels, which set off a safety alarm. According to the tests Gareth would have survived no longer than thirty minutes had he been alive in the locked bag. I used another bag with the end cut out to give me more room so I could see if it was possible to manipulate the lock in any way.

We then took the equipment to a hotel and rented a room so we could repeat the tests in a bath. That was even more impossible. We videoed it. I dread to think what the hotel staff thought when a group of men walked in with two empty red holdalls, camera equipment, padlocks and monitors.

It was a horrible experience.

I read up on ways to get into bags and found out about an old baggage handler technique exposed years ago after a group of baggage handlers were caught using pens to prise zips on bags apart without undoing the locks. They could then reseal them by running the fasteners back along the open zip. It was impossible to do this from the inside of the bag, however.

In Gareth Williams's case there were no fingerprints, no footprints, no DNA. Surely if he had miraculously managed to climb into the bag and lock himself in, there would have been evidence of his struggle? I tried over 300 times and couldn't do it. Not even Houdini would have been able to pull it off.

I presented my thoughts to the investigation team. In my opinion Gareth was already dead when he was put in the bag. It was lifted into the bath because that would have allowed the decomposition fluids to run away, thereby not causing such an overpowering smell that would seep into other flats and alert people. The heating was turned up to hasten decomposition and destroy evidence of injury.

In December, further details about Gareth's private life were released and the death was officially described as 'suspicious' rather than murder. It was explained that before his death he had been to the US West Coast for his work. He was known to be involved in a computer hacking investigation. On 13 August, back in the UK, he went to a drag show in East London on his own. He also had two single tickets for two other drag acts at the Vauxhall Tavern in South London, not far from MI6 headquarters, for each of the following weekends.

One witness said they chatted to him at a gay bar earlier in the year, but police were unable to trace any sexual partners.

Between May 2009 and the time he disappeared, he had visited five separate bondage websites on four occasions which were 'how-to' sites rather than pornographic websites.

E-fits of a Mediterranean couple said to have visited the flat were also released and there was an appeal for anyone who had encountered him in the nightclubs, online or at women's clothing shops to come forward.

Details were also given of the women's designer clothing and shoes found in his flat along with several wigs. All the attire would have fitted Mr Williams, it was claimed. It was also revealed that he attended two, six-to-eight-week courses in fashion design for beginners at the prestigious Central St Martin's College of Art and Design in London during evenings and weekends, one in 2010 and one in 2009, passing both courses. Clearly the Met were actively pursuing the sexual deviancy line. The investigation was leaky. Someone was feeding stories to the media.

I wasn't buying it. Perhaps it was a red herring, or perhaps Gareth's undercover work required him to pose as a woman. I spoke to people I knew in the Secret Service, and they confirmed that it was plausible that his undercover work involved posing as a woman, or as a drag queen. It looked to me like someone was trying to smear him, which was sad, because he was working in the service of his country.

The case rumbled on for nearly two years. I was repeatedly asked for my opinion, and I thought he'd been killed.

In March 2012 I gave evidence at the inquest into the death, which was held by coroner Dr Fiona Wilcox. She asked my opinion. I told the inquest that I wasn't a detective, but Dr Wilcox said I had vast experience and wanted my opinion. I told her that the heating was turned up full, the bathroom door and the shower screen were closed, and it would be impossible for Gareth to do this to himself without leaving a trace of footprints, fingerprints or DNA on the bath, bag, padlock or shower screen. I told the coroner that in my opinion Gareth had been murdered. The family believed this too and were frustrated by the continual focus on his private life. They believed someone was in the flat with him or had broken in afterwards and stolen items. They mouthed a thank you to me after I gave evidence. The evidence to support the claim that Gareth was involved in some form of deadly kinky sex game was weak and Dr Wilcox interrogated it. At one stage when one of the witnesses explained that semen had been found on various pieces of furniture in the flat, she explained that it proved nothing and that there were probably traces of semen on many people's furniture, at which point the room erupted into laughter.

To me it continued to appear that there was a smear campaign going on and I thought it was awful.

Dr Wilcox recorded a verbal verdict that the death was 'unnatural and likely to have been criminally mediated'. She believed that on the balance of probabilities Gareth was killed unlawfully and that someone locked him in the bag and put the bag into the bath.

The verdict pre-empted another Met investigation that lasted a further twelve months, after which the investigators were still convinced Gareth had done it himself. There was also a video which showed a girl, three inches shorter than Gareth, who managed to get inside the same type of bag and padlock herself in from the inside – after a lot of struggling. When this was pointed out to me, I explained that the critical fact in Gareth's case was that even if he could contort himself into the bag and padlock it closed from the inside he would have left a trace of DNA, fingerprints or footprints.

In November 2013 Metropolitan Police Deputy Assistant Commissioner Martin Hewitt announced that despite a re-examination of all evidence there were still no definitive answers and that in his view the most probable scenario was that Gareth died alone in his flat having accidentally locked himself inside the bag.

Two years later the case refused to die. A former KGB agent, Boris Karpichkov, who had defected from Russia, gave an inter-view to say that Russian agents had killed Gareth after trying and failing to blackmail him into becoming a double agent. According to the Russian spy, Gareth had countered their threats by claiming to expose the identity of a Russian agent in GCHQ. It might have sounded fanciful in the days before the Salisbury poisoning, but if recent history tells us anything, it's that Russia is more than capable of carrying out audacious assassinations on foreign soil.

And in 2021 it was reported that advances in DNA testing had made it possible to sample a piece of hair found inside the bag over a decade ago for precious DNA details. Tantalisingly it appears that forensic science may yet finally tell us what happened to the spy in the bag.

Chapter 18

More than anyone, I knew that surface appearances could be deceptive. When you go beneath, things are not always what they seem. The swan gliding serenely on the water has its legs going like the clappers underneath. That well-laid garden path or nice new floor tiles in the kitchen? Who knows what's buried beneath them, what secrets they hide?

That same rule, that what we see on the surface is only half the story, applied to my personal life at the turn of the decade. The company was going from strength to strength; I was comfortable; I had a four-bedroom detached house on Reigate Hill.

Although I liked where we lived, I had always wanted to move further out to the country and when the opportunity arose to purchase thirty acres of land in Sussex, I jumped at the chance, figuring the plot would give me physical and metaphorical space. Ever since I was a child I'd enjoyed being outdoors and spending time in the countryside. I liked animals so I got some rescue pigs to look after on the land. I also put a caravan on it so I could spend more time there. It became my go-to place when I wanted to get away. It was my sanctuary and I started to look into the possibility of building a house on it.

While work was going well, there were problems in my home life and Mandy and I decided to part. Meanwhile the company was expanding in several directions at the same time. In addition to the forensic and protest work, and round-the-clock Fire and Rescue support, the dive team was also becoming busier. The demand was driven by two factors. Firstly, we had a reputation for successfully finding missing persons and evidence, and secondly, continual cuts in police resources meant fewer forces had their own divers. Even some forces that policed the coastline were without divers, which seemed crazy to me, particularly given how difficult sea searches were.

Across the country, there are very few police underwater search units left. Many had been disbanded. So if you got into trouble in a river or lake, or one of your loved ones drowned, unfortunately there was a slim chance that an underwater search unit would be available.

The dive team was being called upon more and more, all over the UK. We were regularly called by Avon and Somerset Police dive team to assist them using our SSS, to rapidly locate missing divers in particular. Usually we used our equipment to find targets and the police divers went down to investigate. Over the years I lost count of how many jobs we did.

One particular wreck, reportedly the most-dived one off the Dorset coast, seemed to claim more lives than others. As far back as 2005 we were being called to search the *Kyarra*, a twin-masted, schooner-rigged, luxury steel steamer that was sunk by a German submarine off Anvil Point in 1918.

Even if a body was located, conditions underwater sometimes meant it was too dangerous to recover it. This happened to one victim of the *Kyarra*, a sport diver who we managed to find lying off the wreck using sonar. Although we located the body, the current was too strong to get a diver in and the search had to be abandoned. Two years later, I got a call from a contact at the police to say that the body had finally been recovered from the same place we had identified. The weights on the dive suit had kept him in place but when he was eventually brought to the surface all that was left inside the dry suit were just bones.

In April 2011 we were back in familiar territory looking for a sixty-one-year-old man off the coast of Lyme Bay, Dorset. He was an experienced recreational diver and had been diving with a friend who raised the alarm. They had been at a depth of twenty-two metres when the victim's dive buddy realized his friend was in trouble and tried to rescue him.

We were on the scene quickly and searched the one-kilometre-square dive site area, but we failed to find him. Sadly, this was an occupational hazard, and not entirely unexpected at sea where, as in the case of the sport diver, even if we located the remains there was no guarantee we could recover the body. There's a reason they call it the cruel sea.

On land it was a different matter. If we had accurate and reliable intel there was a good chance we'd find the target, as in the case of murder victim Kate Prout, where the information came directly from her husband, Adrian, who killed her.

Adrian was a millionaire; they were divorcing and were arguing over the size of the payout Kate was asking for. The last contact she had with the world was at 3.29 p.m. when she called her bank, First Direct.

Adrian owned a pipe-laying business and a commercial pheasant shoot. Kate had accused him of threatening to kill her in the past and was suspicious he was having an affair when she asked for a divorce on 2 November 2007. Adrian was angry at the prospect of selling his £1.2 million farm in Redmarley,

Gloucestershire, to pay the £800,000 settlement she was asking for. He reported his wife missing on 10 November and on 27 November he was arrested on suspicion of murder, then released without charge the following day. In the subsequent months he was rearrested, and the farm was searched. Eventually, in March 2009, now with a new girlfriend who was expecting their child, he was charged with murder. He was found guilty and jailed for life in February 2010 but continued to protest his innocence, claiming he had been the victim of a miscarriage of justice. Friends and family stuck by him. He agreed to have a lie detector test, which he failed, and eventually he admitted to his fiancée that he had strangled his wife following a row and buried her body. He then led police to the area on his farm where he said he had buried her.

Adrian told officers that after he killed Kate, he wrapped her body in a carpet, put her in his car, went to the pub and then later that night buried her in front of a pheasant pen, using one of the posts to hang his jacket on while he dug the shallow grave.

Soon after the confession, me and my team were called in and found her almost four years to the day since she'd been put there.

The following year, I was involved in the biggest search I'd ever been on when we were called in to assist in the hunt for missing April Jones, a five-year-old girl who disappeared from near her home in Powys after being seen getting into a vehicle.

Once again, the helicopter proved useful. I flew to join the search operation while the team followed in vehicles, also bringing boats so we could utilize the full range of equipment at our disposal.

I'd never seen so many people on a search. There were

thousands of people wanting to help, including mountain rescue teams and specialists such as Mandy Chapman and her dog team from the Met Police. The whole town had turned out to assist, along with other volunteers from further afield and hundreds of police officers. The vast operation was being coordinated from a leisure centre and I wasn't able to get in to talk to the PolSA to get a brief as it was so hectic. Instead, I rang him and asked him what he wanted me to do. There was a remote mountainside and a river to search, and he asked me to do some fly-overs at the top of the mountain and then to search the river with sonar. I took the helicopter up, flew over the area and looked down for any possible disposition sites or areas where someone could hide, or a child could get lost. On a remote path I saw a drain cover that had been removed and left aside. It set alarm bells ringing. I landed on the track and shut the rotors down to take a closer look, but there was nothing in the drain. Some of the public working on the search may have found it, lifted it and forgotten to replace it.

Later that day I went up in the Dyfed-Powys Police helicopter and while flying over the hills we spotted a caravan in a remote part of the woods, which looked suspicious. It would have been an ideal location to hide a kidnap victim. We landed back at the local leisure centre and me and the team drove to the bottom of the mountain in which we'd spotted the caravan and set off on foot to check it. By that point it was around 11 p.m. and the mist and rain had swept in. We hiked for around an hour until we approached the caravan. There was an ominous, creepy silence as I opened the door carefully and ventured inside. It was full of speakers and amps. It didn't look like anyone had touched anything inside it for a while. We took the speakers out to search the space thoroughly, then checked underneath the caravan and put it all back neatly as we'd found it.

The search went on for weeks. The prime minister at the time, David Cameron, issued an appeal for information.

A suspect, Mark Bridger, used to go up into the forest to practise bushcraft. He was arrested within twenty-four hours of April's disappearance and on 6 October he was charged with child abduction, murder and attempting to pervert the course of justice. Two days later, he was charged with the unlawful concealment and disposal of a body. He was tried and found guilty in May 2013. April's body was never recovered but fragments of human bone were found in the fireplace of his home. It was believed he scattered her remains across the countryside near his home.

Life was busy. There was little time for anything else and if I'm honest, that was fine by me because as I hit fifty, I was on my own with no one to share my daily life with. I wasn't looking for anyone, had been single for over a year and wasn't particularly interested in getting involved in the dating game or anything like that. I was not the Tinder type. But then I met Adele, and everything changed.

My friend Jason was the emergency planning officer at the Royal Surrey Hospital and, realizing that I was turning into a bit of a hermit where the opposite sex was concerned, he invited me out one evening to a work get-together. I walked in and immediately saw a pretty blonde girl by the bar laughing and talking to the group Jason was with. He saw me and beckoned me over. I said hello to the people in the group that I already knew, and he introduced the girl who had caught my eye.

'This is Adele,' he said, 'she likes boats too, you two will get on.'

We started chatting and she was as bubbly and friendly as she looked. She explained that she was the communications manager at the hospital. We chatted for a while and then she left, explaining that she had a prior engagement she needed to get to, which secretly I was gutted about, but we said maybe we would catch up again and left it at that.

A few weeks later Jason was having a 5 November party for which I agreed to build the bonfire and I took the search trailer from work along to use as a bar. It still had all the axes and cutting equipment hanging up neatly inside and looked quite impressive. Mum and Dad came along too and as I was preparing the fire, covered in dirt and stinking of smoke, a familiar face walked over to say hello. It was Adele and, being the gentleman that I am, I led her inside the trailer bar to pour her a drink.

While the Merlot was pouring, I could see her looking around.

'What is this trailer we're in?'

'It's for my work,' I replied.

'Are you an axe murderer?' she joked, not having a clue what I did for a living.

We carried on where we'd left off and got on like a house on fire, which was ironic given the date. At the end of the evening, I asked if she fancied meeting up for a drink, and over the following months we met up a few times as friends. We enjoyed each other's company and had a lot in common. Like me, she had recently gone through a marriage split. She was Canadian, a sailor, loved the outdoors and was close to her family. I was hooked. Just before Christmas, we became a couple. It was like having a new lease of life, we got on so well, it was like we'd known each other forever.

In 2012, I proposed to Adele during a romantic dinner on a beach in the Grenadines and was overjoyed when she said yes.

We also decided to start a family together and after a gruelling round of IVF, she fell pregnant. The prospect of being a father again was exhilarating. I couldn't have been happier. We rented a small house and made plans to buy a permanent home in the countryside with some land where we could keep animals and raise our daughter to love the outdoors like we did.

In August 2012 we flew to Vancouver, Canada, and got married in an intimate ceremony on a small island with my mum and dad, Adele's parents and her two sisters, Rachelle and Collette.

For our honeymoon we chartered a fifty-foot yacht and cruised around the magnificent Desolation Sound. On the flight up to Comox, where we were collecting the boat, Mum's metal hip set off the alarm through security. The security guard asked her to open her bag, which was padlocked, and she couldn't find the key.

'Has anyone got a hairclip?' I asked.

I reached over and pulled one from Mum's hair then kneeled down and deftly fiddled the lock to open her hand luggage, using skills I'd learned to free protesters from lock-ons during count-less protest jobs.

Collette was with us and watched in amazement.

'Wow,' she said, 'that is so cool.'

From then on and to this day forward, her nickname for me has always been 'Bond'.

It was an idyllic trip, being on this beautiful boat, in this beautiful part of the world, happily married once again, expecting a baby, and thankful that Dad was with us. He was ill at the time with prostate cancer but was in good spirits and loved the boat and the incredible scenery. It was the start of a new chapter in my life and I couldn't be happier.

*

Many of the cases I was called to were tragic. It was hard not to become emotionally involved. On forensic searches, I could hide behind the armour of cold objectivity. I was there to do a job and in order to do it to the best of my ability, I needed to be scientific and methodical. There was no room for emotion. But when children and young people were involved, it was impossible not to feel sad, or angry.

The worst for me were the drownings. Not because they were any more tragic than the murders, it was the senselessness and the preventability of them that got me. In the cases where we did recover the body, we literally had direct contact with the victim. In many murder cases we didn't, we were there to look for evidence.

In August 2013 we were called to recover the body of a twenty-year-old man whose death was entirely avoidable. He had been drinking and waded into the River Thames near Hampton to swim across the river after he missed the last ferry home. He was wearing a rucksack. Witnesses said he made it two-thirds of the way over then sank under the surface. Two schoolboys who witnessed it jumped in and tried to help him but it was too late. We were called quickly, and it took us three hours to find him. He was in two metres of water.

Just one silly decision and his life had ended.

That year, 2013, was a busy year for me personally, firstly with the birth of our daughter, Summer, and then finding our forever home, a country estate in West Sussex. With a lot of work, imagination and money, we knew it could be perfect. It had a lot of land, a few lakes and several woods. There was a main house, outbuildings where I could store the helicopter and other equipment and a separate cottage. I was also toying with the idea of hosting forensic search training sessions and would need somewhere with land on which to hold demonstrations. The place had no shortage of space, the whole property was 140 acres.

I realize it sounds very grand, but there was one problem. The main house was almost derelict and the land was completely overgrown. The house had been vacant for four years, the ceiling had collapsed in the hall, the windows were rotten and falling out and the inside walls were black with damp. The driveway was covered in thick moss and weeds. You couldn't see the lakes because the grass and brambles were so high. The cottage on the site was also derelict and the barn was full of rubbish.

But Adele and I were not people to shy away from a challenge and we went to view it. When the massive iron front gates first opened to let us in, she said: 'These gates are opening for us.'

I wanted a proper look round but the land was so vast and so overgrown it was impossible to walk around on foot, so one of the lads from work brought over our Argocat eight-wheeler multi-terrain vehicle on a trailer.

The place was a repossession and was owned by the bank, which was planning to parcel it up and sell it in lots. But Adele and I saw the potential of the whole site. We could keep animals and my parents could live in the cottage in the grounds.

In the end it didn't take much thought. It was a massive stretch financially and the mortgage lender would take some persuasion, and then there was the work involved to fix it up. It would take a huge effort to clear the land and to get it back to a basic live-able standard but I put an offer in for the whole site. There followed months and months of surveys and negotiations with different lenders.

While all this was going on Dad's condition was getting worse.

When he was fifty-eight, he was extremely fit and well. One day, he told me he'd had his medical MOT and had been put on blood pressure medication and statins by the doctor. Within months of going on medication he became tired and a different person. He started to limp and then needed a new knee. It was frustrating for him because he was so fit and active, and it was tough for me to witness the man who was such a colossus in my life and who had been so influential become so frail. The glint in his eye and his sense of humour were still there but his body was letting him down. By late 2013 he was eighty-three and living in a hospital bed in the front room of the house. Then one day, when he was particularly weak and struggling to breathe, he was taken into hospital.

I took Mum to see him and was shocked to see him in so much distress. He was gasping for breath and was on oxygen but not on a drip. I'm not a doctor but I did have some medical experience and I could see Dad was dehydrated and needed someone to check his respiration. I asked to speak to a doctor. I was told none were available for three hours.

'My dad needs urgent fluids and he needs a respiratory doctor,' I said. 'Surely someone can put a drip in him?'

'Are you a doctor?' came the reply.

I was getting agitated.

'I need a respiratory doctor to come now and see my dad,' I repeated.

Again, I was told no one was available as it was the weekend.

For the next three hours I sat with Dad while he continued to fight for every breath. Eventually a specialist respiratory nurse came along and agreed that he should have been on a drip. It transpired that he had been on a drip but it had been taken off.

The nurse went off to check why this was the case. The patients were being looked after by healthcare assistants. I know it wasn't their fault – they were so short-staffed.

I willed him to push forward and to concentrate on slow and deep breaths. He had been the guiding influence in everything I had done. He had taught me so much and supported me throughout my life and now I was there supporting him. I sat quietly with him, reflecting on the adventures we'd had together. Dad meant everything to me.

By the time he was seen and made more comfortable, it was midnight. I asked the respiratory nurse what was going to happen, and she explained that Dad would be put on antibiotics and should be out in a few days. Mum and I were exhausted so we left for home to get some sleep.

The following morning, Sunday 17 November, Mum rang the ward to check on Dad. She called me to say he was okay; he'd been comfortable all night. I put the phone down breathing a sigh of relief. Ten minutes later, the phone rang again.

It was Mum crying and wailing. The words barely registered. 'Dad has died.'

I couldn't speak. I didn't know what to say. Time stood still. The phone, pressed to my ear, was silent for what seemed like hours.

'But they said Dad was fine,' I stuttered.

Looking back, I think that possibly Dad had been dead for hours and nobody noticed. It was only after Mum's call that they did a proper check.

The next hours and days were a blur. I repeatedly broke down in tears, but I tried to keep it together for Mum's sake because she was in pieces. I went to the hospital to see Dad. I put on a shirt and tie and a three-piece suit to show respect. Dad had always taught me to be well turned-out. It was what he'd expect.

He lay in the bed where I'd left him the night before. He looked peaceful but drawn. I stood by the side of the bed sobbing. A hole had been torn in my life and it was never going to be filled in.

I got back home that afternoon and was in a daze, trying to process what was happening when there was a knock at the door. It was the landlord of the house we were renting informing us he was selling up and wanted us out by Friday.

'My dad has just passed away,' I told him.

'I'm sorry to hear that, but I need you out by the end of the week,' he reiterated.

Over the following five days I had to organize the funeral, deal with the death certificate, find somewhere to live, pack up the house, run my business, care for a newborn and my grieving mum, who was also moving in with us. It was just carnage. My head was all over the place.

Thankfully the team at work stepped up and took some of the pressure off. Jane, who works in the office, was brilliant and found a house for us as we were still negotiating for a mortgage.

We had a family burial plot and, because of what I'd seen in the line of duty, I wanted my dad buried in a copper casket, not a wooden thing that would rot. I made all the arrangements and before the funeral my team helped me move everything into the new place, which was a short-term lease.

For a few weeks things were mercifully quiet and we were left to grieve and come to terms with our loss. But nature continued unabated and had plans for us. The day before Christmas Eve the seasonal storms arrived and our contract to provide support to Surrey Fire and Rescue came into its own.

It had been a wet winter, with continual weather fronts sweeping in from the Atlantic, dousing the country with rainfall. The South East was hit particularly hard and from the SGI base in Dorking I'd been watching the River Mole rising. I was concerned and I told the Emergency Planning Department at Surrey Fire and Rescue that I thought the river was going to burst its banks. We got on a conference call with the Environment Agency and were told the computer modelling was saying there would not be flooding. We agreed to regroup in the morning but I was not happy. I'd been measuring levels at different points along the river, and it was dangerously high, the flow was not slowing and there was more rain on the way.

I went back to the unit and told the team to be ready.

Sure enough, at three the next morning my pager went off. Outside, the most horrendous storm was bending the tops of the trees. I called the control room and was told there was an incident in Cobham where a tree had come down onto the roof of a house and there was a person trapped inside. I drove on blue lights, dodging branches that were coming down around me. From the unit in Dorking, our response vehicle with its chainsaws and rescue equipment was scrambled and we both arrived at the scene together. The tree had practically cut the house in half and luckily all the people in it had got out safely and no one was injured. We had just started back when the pagers went off again and we got diverted to the Burford Bridge Hotel at the base of Box Hill, on the banks of the River Mole. The report said thirty-two residents were trapped inside by rising floodwater.

At the time, firefighters were due to hold a strike but they decided not to as the major incident unfolded. When we arrived at the hotel the waters were still rising and the Fire Service was there, helping residents wade to safety.

We helped too until another emergency call came through. A man was trapped on a milk float amid rising floodwater down in Sidlow. Across the county the rivers were bursting their banks. The milk-float job was deemed more urgent, so we raced to the scene. When we arrived at the location we had been given there was no trace of anyone or anything so we went back to the hotel and carried on getting the residents out. Another call came in confirming the location of the milk float. It was near the big old country house that my mum used to work in as a housekeeper, so I knew the area well. We raced there and found a man stuck on the top of a milk float in racing floodwater – you could only see the top few inches of the vehicle he was on. He was shaking and petrified. We launched the boats and went out to get him. My team literally dragged him from the top of the milk float into the boat and got him to safety. Soon after the vehicle was completely submerged, and he would undoubtedly have drowned.

As we were taking him to the hospital to get checked over another call came in. A young girl, her mum and a man were trapped in a taxi in rising water near Brockham Bridge. We got to the scene and firefighters were there but couldn't get to them because the water was flowing too fast. The people had climbed out of the car via the sunroof and were on its roof. My team put the boat in and worked in the fast current to get to them. It was a dramatic rescue, and they were pulled to safety just in time as two minutes later the car turned turtle, went upside down and was washed away. The girl, her mum and driver would have drowned. Later, four of my team received bravery awards for that particular rescue.

There was no let-up at the time, however, and we were called straight after to help trapped residents at a mill in Dorking.

The pressure was on constantly for all the responders that day. We went from job to job, plucking people out of the water from homes and vehicles. We'd been in dry suits for hours. It was non-stop and exhausting, but we were saving lives and that was what mattered.

I'd never seen anything like it. I had predicted it was going to happen but no one would believe me – and that was the disappointing bit. We rely on computer models for everything but unfortunately they are not always right. Sometimes experience and human intuition are better than algorithms.

Being a pilot, I knew the implications of the weather fronts that were coming in before the floods and I'd been out and about in the fields and countryside enough to know that the ground was completely saturated and could hold no more water. There was only one place for the rain to go: into the rivers.

We helped out for three days over Christmas. We recovered a gentleman from the upstairs of his house who had just had a hip transplant. He had no electricity or heating and the downstairs was flooded with muddy water. I contacted fire control and requested the address of the local council refuge that had been set up during the floods.

'Unfortunately, it has closed for Christmas,' the operator said. I was fuming at this poor decision when people were suffering. I asked the gentleman where his nearest family was. He said they were in Maidstone so I instructed two of my team to drive the man in our vehicle so he could be with his family for Christmas. We finally finished at 3 a.m. on Christmas morning.

Buildings were submerged, cars were underwater. The area had never seen anything like it. In places it was biblical.

The waters finally receded but the storm fronts continued and in February 2014 it happened again. This time the Thames flooded in Staines and Chertsey.

We helped multiple people trapped in a housing estate and got them to safety in the boats. It was such a big operation, everyone helped – the Army, the Fire Service, paramedics, the Police, volunteers, the Salvation Army.

We waded into one house and there was a very large man in the lounge sitting there with water up to his shins.

'Hello sir, how are you?' I asked.

'I'm alright and glad to see you,' he said.

'Don't you worry, we'll get you out shortly. We have a boat outside,' I told him.

He looked down at himself and then at me.

'You're going to need an effing aircraft carrier for me, son,' he laughed.

He was a real character and a lovely man and after a long day, a small moment of laughter like that was a tonic for everyone.

The whole job lasted days and we utilized our helicopter to provide aerial photographic support using a military photographer on board. The pictures were used to update the daily COBRA meetings in London. The waters finally subsided and the inquest into why it was so severe started. Personally, I've seen adverse weather events come and go over the years and although undoubtedly climate change is happening, all the defences in the world would not have stopped what happened at that time. The water was coming from all directions.

The following year, 2014, started with gathering storms, but we had something to celebrate in the spring when we finally got the mortgage approved for our new home. Unfortunately, Dad never saw it. We moved in April and one of the first jobs was to renovate the cottage on site for Mum to live in so she could be near us. It was a bittersweet welcome to our new home, with a new baby, Mum in her cottage, but Dad no longer with us.

Chapter 19

I threw myself into work. I guess it is my coping mechanism. Luckily, we continued to be busy and the jobs – as varied as ever – continued to roll in. Dad's loss left a huge hole in my life. Thankfully I had Adele and Summer, Natasha and Danielle and we all supported each other and looked out for Mum. We all had good days and bad days and often I'd be on a job and would have to take a moment to gather myself because something would happen that reminded me of Dad. I'd find myself doing something that Dad had taught me and I'd silently thank him.

In 2015 the protesters were at it again, which gave me plenty of opportunities to reflect on the jobs that Dad and I had been on together and to come to terms with his passing. On one job an industrious group had set up camp in an allotment in Bristol, amusingly calling their campaign Save our Celery. Dad would have laughed at that. The site was due to become part of the city's Metrobus scheme and I flew over the area in the helicopter to get an idea of the scale of the encampment. The eco-warriors were well-embedded, so after two weeks of careful planning we went in at first light with bailiffs, who started sealing off the area. There were also security guards there and they told me that they first planned to tow away a big old caravan that was blocking the entrance and impeding the eviction. I immediately stopped them.

'Nothing gets moved until I say it's safe to do so,' I said.

It was lucky I intervened.

On the door of the caravan was a poster with a picture of a tree house attached to a cable attached to a caravan. The words 'danger of death' had been written underneath. The warning applied to a hanging tree house which was held in place by a cable secured to the caravan. If the cable was cut to move the caravan, the tree house, which had people in it, would come crashing down to the ground.

The caravan was locked but with some crafty lock picking I was swiftly inside where I found a duvet on the floor. Underneath was a girl lying on her side with her arm reaching through a hole in the caravan floor. I went outside and looked under the caravan to see where her arm went. I was horrified to see it was locked into a tube that was concreted in the ground underneath. I gave silent thanks that the security guards had not hitched the caravan to their 4x4 and towed it away – the girl's arm would have been ripped off.

We dealt with the cable and tree house first, then a guy who was sitting in a hole in the ground chanting to himself and was locked into a pipe concreted into the ground. Finally, we got the caravan girl. That was a tricky one, as we had to crawl under the caravan to dig round the pipe and release her.

The rest of the operation was complex. It took four days to remove protesters from tree houses in tall poplar trees. When the job was complete, I had to marvel at how, after all the years, protesters still managed to surprise me with their ingenuity, and the multiple ways they were willing to harm themselves for their cause.

Over the following years there were more significant protests against fossil fuel exploration. At one, in Leith Hill in Surrey, police were rightly reluctant to commit lots of resources, unless it became a public order issue. It was my contention protesters should not be arrested and charged because the burden then gets shifted onto the overstretched courts, and for what? A small fine and a slap on the wrist.

In a way, Leith Hill was like a 'greatest hits' of previous protests. There were tunnels, tree houses, a tower, barrels filled with concrete and tar, a lady locked into a 4x4 bull bar by the neck using a kryptonite D-lock.

We cut our way into the fortress using reciprocating saws. It was an impressive structure complete with a spotless kitchen. We cleared the camp methodically and safely. The Fire Brigade was on standby in case the pallets were set alight.

Later that year we were involved in the removal of anti-fracking protesters from a camp near Horse Hill, Surrey, where UK Oil and Gas Plc had started drilling for some of the estimated 124 billion barrels of oil under the Weald Basin.

In between all this I found time to expand my repertoire of skills and went to the US for seven weeks to qualify for my US pilot's licence on a Cirrus SR22 Turbo, a high-performance aircraft.

The range of work kept me occupied and gave me a focus away from the grief that I was still dealing with. I was glad to be so busy.

Back in the UK, I found myself at a murder scene in a sleepy Surrey village, looking for evidence after a fatal shooting. The circumstances of the shooting were highly incongruous to the location. Headley is a tiny, quiet village with a single pub and post office, surrounded by rolling hills and horse stud farms. The shooting, however, happened during a massive party in a rented mansion full of over 400 reggae fans from Brixton and South London. Early reports, denied by the organizer, said that it was a swingers' party. Ironically, it was held a stone's throw from the appropriately named local pub, The Cock Inn.

Pictures from the party showed chaotic scenes of champagne-swilling guests packed into the pool at the house, the girls in skimpy thong bikinis and high heels. The sound system pumped out music until the early hours when the bash descended into mayhem. Shots rang out. One guest, Ricardo Hunter, was killed and a woman was shot in the leg.

The following morning, with the detritus of the party still littering the property and surrounding road, we got called by the police search adviser, primarily to do water searches. When we arrived, the whole village had been cordoned off with police tape.

The first area we were asked to look at was a small pond outside the local church.

'I'll dive,' I joked. It was twelve inches deep.

Instead, I used the underwater metal detector and lay down in a carpet of pond weed in my dry suit, searching through the weeds potentially for a handgun and ammunition. I didn't find the weapon, but I did find the hard drive for a computer, which seemed like a strange thing to dump in a pond right outside a church.

While we searched the pond, police teams were looking through the surrounding undergrowth. There was also a major forensics operation going on inside the house, where we headed once we'd cleared the pond.

There, we were tasked with searching the pool. Although the water was clear, it was full of the leftovers from the party. There were lots of clothes and bottles, and heaven knows what else floating around. The pool was contaminated with urine and other possible bodily fluids so I chose to wear my dry suit.

It was a heated pool, and the work was sapping. I felt like I was being poached as I meticulously moved along the bottom, looking for a gun, or a cartridge or anything else of interest. Again, we drew a blank, but at least the investigation could then tick the two areas of water off the list of possible disposition sites.

As of December 2021, when an inquest was held into the shooting, the killer or killers remained at large. Suspects had been identified and there had been a *Crimewatch* appeal, but detectives told the inquest that they had failed to turn up any solid evidence and that witnesses had been uncooperative.

Evidence recovery was a big part of my work for the police. Over the years I had recovered many guns, knives, samurai swords and even stolen ATMs. In one case, in West Yorkshire, two drug dealers had been shot in a lay-by. When the killer was apprehended, he admitted throwing the weapon, a twelve-bore shotgun, into a nearby lake. As the usual underwater search team was not on duty that weekend, we were called in. We found the gun with sonar and an underwater metal detector.

The shooter claimed the killing was an accident and that he had discharged both barrels when he slipped. However, when we found the weapon and opened the breach, only one cartridge was still in the barrel. In a twelve-bore shotgun, once the weapon is fired, the cartridge will remain unless the breach is opened for it to come out. For his story to be true, two cartridges would still be in there. The single cartridge meant that he had shot one round, reloaded and then shot again, so it was far from accidental.

Later in the summer of 2016 we were tasked with looking for

evidence at the scene of a murder in which an innocent angler, who had been camping on an island in the River Thames near Walton, was beaten to death by three thugs who were also camping on the island.

Scott Wilkinson often went on fishing trips to Donkey Island near Sunbury Lock and would sometimes stay there for several days at a time. His girlfriend and friends kept him supplied with food and clean clothes while he was on these expeditions. In July he was on the island when brothers Lenny and Shane Crawt, both teenagers, and their cousin, twenty-one-year-old Charlie Smith, turned up to camp there too. Lenny, who was just sixteen and originally from the area, was on the run from a children's home in Blackpool and was hiding out on the island to avoid detection.

On 27 July at 11.18 p.m., a heart monitor implanted in Scott's chest recorded a sudden, erratic heartbeat as if he had suddenly received a spike of adrenaline. At around the same time, a witness living near the island said he heard several male voices shouting. At 11.22 p.m., CCTV showed the three boys leaving the island. By that point, Scott's fate was already sealed. The boys had attacked him with a blunt object, and he sustained catastrophic head wounds that irreversibly damaged his brain. He remained alive for another three hours. His body was discovered by the water's edge the following afternoon. It was never established what led to the attack.

We were tasked with searching the river for evidence and I was the diver on duty that day. When I heard about how sense-less the killing had been, and that Scott was just a man on his own enjoying the outdoors, I was determined to fingertip search every inch of the riverbed around the murder scene.

When I looked around Scott's camp, I noticed that a branch

had been cut from one of the trees he was sheltering under. I recognized axe marks and there were wood chips underneath. His fishing rod rests had also been drilled into a tree. I made a mental note to look for an axe and a drill, as to my knowledge none had been recovered.

I was searching underwater for two hours and twelve minutes in the end, feeling around, hoping to find something that might nail the killers. It was the longest any of my dive team had ever stayed down and I was determined to find the weapon. On my very last sweep I found a large kitchen knife. It was taken away and formed part of the evidence against the men.

In March 2019, the Crawt brothers, who were under eighteen when they killed Scott, were found guilty of murder and sentenced to twelve years each. Smith, who was older, was found guilty of manslaughter and charged with thirteen years.

We were repeatedly called to the water that year, 2016. It was one of the worst years I can remember for drownings. Again and again, we were pulling up bodies from rivers and lakes.

On one job, some young Polish men swam off a friend's private boat dock in the River Thames and one of them got into difficulty and disappeared below the surface. When we arrived, we were told by the firefighters, who had received their information from members of the public, where to search. I insisted on speaking directly to the family to get an exact location and was given a totally different area to search.

Our first diver on that job was Barney. He dived for forty-five minutes on a specific search pattern while I sat as the standby diver fully kitted up, having cold water poured on my head to keep me cool. It was then my turn to dive and I picked up where Barney left off in the murky water, swimming slowly along the underwater cable, or jackstay, which I then moved across one metre and started a new search line. As I swam along feeling around in the dark, I suddenly felt the shoulder of the drowned man. Only one hour before this the man had been enjoying an afternoon with his friends on the river and now he was another unfortunate statistic. With his eyes wide open staring at me, I held him tight under my arm and swam him back to the dock.

Heatwaves in July and August meant people flocked to the water to cool down and enjoy the sunshine. Sun, water and alcohol repeatedly proved a deadly cocktail. In the River Medway a young lad had been drinking on a bench by the river and decided to swim across. He got in trouble and drowned in the middle. We recovered his body soon after.

In Virginia Water, a beauty spot in Surrey, a young guy was on his first date with a girl on a warm Friday evening and they decided to take a dip in the lake. They swam out but he got tired and started to struggle. She was a strong swimmer and tried holding him up, but she couldn't and he went straight to the bottom in about twenty-four feet of water.

She raised the alarm and the Fire Brigade attended first but couldn't locate him. It was dark by the time we got there. The fire boat was anchored where the girl reported that he went down. We put the SSS in and within a few passes the image of a body lying on the bottom, curled up as if asleep, appeared on the screen. My divers went down to bring him up.

A few days later we were back in the water again, in the Thames near Sunbury Lock, looking for the body of a fifteen-year-old boy. We recovered him in the afternoon. His distraught relatives were there waiting.

Speaking to the loved ones and the friends or families who

had witnessed drownings was something I have done a lot of over the years. Each case was individual, and it is one of the hardest parts of the job. It is also one of the most necessary, as witnesses can help locate bodies.

One of the toughest tasks I've ever had was to talk to a nine-year-old boy whose father had fallen off the boat they were fishing from and drowned. The poor lad was alone and adrift in the boat, calling for his father, before someone rescued him. We were called the following day to help with the police operation and I asked to speak to the son because he was the only one who knew where his dad had fallen in.

He was accompanied by a member of his family and a police officer when I met him by the water's edge.

'I'm here to find your daddy. I'm really sorry, but can you tell me where he fell in?' I asked gently.

He pointed to a spot in the river. I thanked him and had to walk away. I didn't know what to say. There was nothing I could say to make his situation any easier but at least I could find his dad for him.

I briefed my team and the divers dressed and got ready to go. With the information I received from the little boy and sonar, I knew I would find his Dad very quickly. And I did. The familiar shape appeared on the screen. We put in a weight, attached to a line and buoy – a shot line – to mark the location.

The family was on the water's edge, waiting for news. The recovery had to be handled carefully and sensitively. We needed somewhere secluded and private to 'land' the body. When there were members of the public around, we put the bodies in body bags underwater and if possible, would swim the remains to a secluded location before taking them out of the water.

Once the shot line was in, I went back and briefed my team and the police. I wanted to explain to the family what was going on because no matter how awful the situation was, they needed to be kept informed. A police officer took me to the family to provide the harrowing news – mercifully the boy had been taken home.

'I'm afraid I've got some bad news,' I said. 'Unfortunately, I think we have found him.' The wife and the mother of the man both collapsed in front of me, screaming in grief, shock and sorrow. It was a very difficult and emotional moment. I offered my condolences, then I had to turn and walk away, wiping tears from my eyes.

I had always understood about grief and how raw and painful it is. I felt it first when I lost my nan. The grief I felt when I lost my dad was so intense and stays with me to this day. Everyone's grief is different but I could empathize with how this family were feeling and I knew from bitter experience that there was nothing anyone could say or do that would make it any better.

Along with the team, I then got on with the grim task of recovering the body. Underwater the divers swam him a little further down the river where there was privacy. I followed in the boat and handed them the body bag. They put him in it underwater, out of sight and then we lifted him out. The family never saw a thing.

In summer, rivers become conveyor belts of death. Weirs are particularly deadly. One, in Reading, claimed two lives within two weeks of each other. In the first incident the victim was walking along the top of the weir, slipped down it and got tumbled over repeatedly at the bottom and then spat out, dead. It was a hot day and we recovered him quickly and landed the body into a tent we'd set up. The flies were on him immediately, even when he was in a bag. We were told the undertakers couldn't take the body away for a couple of hours and I had to make the decision to get the body back into the water to keep it cool and keep the flies away. We put it on a stretcher on a secure line out of the way of the public.

The undertakers eventually arrived and put the body in their refrigerated van. As they drove off the rear door flew open, and the body spilled out on a slider. We chased after them down the track to push the poor guy back in.

Two weeks later we were called back to that place to recover the body of a teenager who walked across the weir and suffered the same horrible death.

When we found a body or a target, we used a 'five bells' code. If the diver was underwater and found something he would tug his lifeline five times so the person on the other end at the surface knew the target had been located. If we were on the surface and there were members of the public around, or if we were using the radio and people could overhear, we used the words 'five bells'.

That summer we attended sixteen drownings. It was constant, and I was continually frustrated by the senseless loss of young lives. All of them were avoidable.

In one case in Guildford, we were called almost immediately after a man fell into the River Wey from the towpath and didn't surface. We found him quickly; within an hour of him going in, and attempted resuscitation. There is a ninety-minute window of opportunity after someone drowns in which it is possible to revive them, but in this case we were unsuccessful.

In one particularly poignant case we were called to a lake near Ripley where a New Age traveller had gone swimming and disappeared. When we arrived, his dog was sitting on the bank staring at a spot in the lake, crying. We put the diver in following the dog's gaze. The water was crystal clear but all we could see from the surface were reeds. Our diver soon found him, though, and brushed the vegetation aside to reveal the body. The faithful pet had marked the exact spot where its owner was.

Of everything that happened that year, though, the most pivotal event occurred in the spring, in early May, when the South East was enjoying a warm spell with temperatures in the mid-twenties.

We'd just enjoyed a glorious first weekend of the month and I was heading to work for a Monday morning meeting in a suit and tie. Adele called the office and asked them to speak to me urgently, she sent through a link to a news story about a family desperately trying to find their teenage son who had drowned. Adele asked if there was anything I could do to help them.

The teenager was Ellis Downes. He was sixteen and had been with his friends on Saturday afternoon by the Thames at Culham, in Oxfordshire. He'd joined them for a dip in the river and became exhausted. They tried to save him, but he went under and never resurfaced. Frantic searches were conducted on the banks and the surface, but poor Ellis was lying on the bottom of the river and could only have been recovered by divers. It was too deep for any other recovery method. The search operation was scaled back quite quickly, leaving the family with no information, answers, or solution. The desperate family took to social media in a plea for help, which was picked up by the media.

At the time, SGI had a retained contract with Surrey Fire and

Rescue so I did the decent thing and rang the chief there, Russell Pearson, to ask if he minded me redeploying the team to Thames Valley to assist a family to recover a body. He gave us his blessing. One of my team then spoke to the Thames Valley and Hampshire Police joint control room to inform them that my dive team was going to help assist with the search. I gave them our details then called through to the SGI control room and briefed the team. My team loaded the equipment and set off. I drove my own car and we agreed to rendezvous at Culham.

As I was on the way the search adviser at the scene called me. He was a PC.

'You are to turn around and go back immediately, you are not to come to this site,' he said.

I was taken aback.

'On what grounds?' I asked.

He wouldn't give an answer.

I told him I'd sort it out with his superintendent when I got there. He told me the superintendent would be too busy to speak to me.

'I'm sorry, but I am going to continue,' I said. He paused for a moment and then said that my team and I were to redirect to a car park in Abingdon away from the search scene where he would meet us.

'Fine,' I said, before hanging up.

It's fair to say his attitude got my back up. I was going to help the family, knowing through bitter experience that no divers would have been deployed because Thames Valley Police did not have a dive unit. It was disbanded in 2014 by their police and crime commissioner as part of wide-ranging cuts in the force.

I called my team.

'Just to warn you, gents, we're going to get some aggro on this one.'

When we got to the car park, the PC was waiting in an unmarked white van. Despite his low rank he was the PolSA, so he was advising on the search. He walked over and told me to turn around and go back. I repeated my insistence that I wanted to speak to his superintendent.

'I can't have amateurs diving,' he said.

'We're a professional dive team,' I corrected. 'I'm going down the river to speak to the family and tell them that we are here to help them if they want us to.'

'Go near the river and I will have you arrested. Do you understand?' he warned. I was getting really irritated with this bloke.

In order to prove our credentials, I phoned the NCA help desk and handed my phone to the PC. They confirmed that SGI was on their database, were a police-authorized team and were held in high esteem. Indeed, we were the only ones in the country who could do police diving work and had full forensic capability. The contact at the NCA recommended we be used.

The PC walked off then and made a call. Ten minutes later he came back, told us we could not dive and ordered us to leave.

'What's your game?' I said to him in disbelief.

'I know who SGI is, I'm an ex-police diver,' he said.

'So that's what this is all about,' I responded. 'There's a lad on the bottom of the river and you are letting politics get in the way of his recovery? I've had enough of your attitude. I'm going to go down and meet the family. If you don't let us dive this is going to go all over the national media by the morning.'

'Bring it on,' he said. 'If you go down there, I'll have you arrested.'

I got in the car and drove off.

I parked just outside the scene, which was cordoned off, and I walked towards the tape. Another officer was standing there who had clearly been contacted by the PC and warned of my arrival.

'Mr Faulding, you go any further and I'll arrest you. This is a crime scene,' he said.

'It's a crime scene, is it? Okay. Show me your crime scene log please, officer,' I requested. 'You are messing with the wrong person,' I added.

'Erm, we don't have one,' he stammered.

'Because it's not a crime scene, is it? You're lying to me,' I said. 'I'm going to walk under the tape. If you want to arrest me, please carry on. Now, excuse me.'

I lifted the tape and carried on walking; I knew I wasn't breaking a law.

The officer did too and stood there not knowing what to do or say.

At the scene I introduced myself to the distraught family and looked around. There were masses of the public along the river-bank. There were a few police community support officers, and no search activity was going on. A reporter from a national newspaper was talking to one of the friends of the family on the phone following up from the family's online plea for help. They were appalled to learn that we were being stopped by the police from searching for Ellis.

The search was the most poorly conducted operation that I had ever seen, with a total lack of communication and coord-ination. The initial search had been carried out by firefighters and volunteer rescue teams who did the best they could with the equipment they had, such as pole cameras, which are ineffective in deep water, and I commend them for their efforts. There had been sonar searches on Saturday night and Sunday, but the equipment used was not effective; if it had been, Ellis would have been found immediately. The Environment Agency had also sent out a boat in an attempt to find Ellis.

I tried to find someone in authority to explain that I had a dive team and that the family wanted us to help.

No officer would come over to speak to me or the family.

The story was gaining momentum in the news by this time and more and more members of the local community showed up on the riverbank to offer their support.

Eventually an inspector turned up that evening and the family surrounded his car. They were angry and they wanted answers. Shortly after a critical incident was declared, which basically meant it was all going pear-shaped.

Ellis's dad, Darren, was desperate for us to get in the water to find his son. He was sobbing, pleading with the police to let us dive. I was horrified by what I was witnessing. In the end the family said if the police would not allow us to go in, they would go in themselves. They'd already hired a boat to look.

By the time the inspector finally capitulated it was getting dark. We didn't have floodlights on board because we had assumed when we set off that morning that it would all be over quickly, and we would be home in time for tea. I asked if we could borrow floodlights from the police but were told they didn't lend flood-lights to 'civilians'. It beggared belief how deliberately unhelpful they were to prove a point rather than act for the good of a desperate family.

Members of the public carried our boat and dive equipment

to the water's edge, while the police officers and PCSOs stood back and watched, not offering any assistance whatsoever. I was absolutely disgusted.

I went on the boat and ran the SSS over the area where Ellis's friends said he went down and found his body straight away. He'd been there for three days. My team dived and recovered his body. It was all over in forty minutes.

It's standard practice when bodies are recovered from water to take a sample of the water the body is found in, in case it is needed for forensic tests. We asked the police for a water sample bottle. They said they didn't have one available, so we emptied a mineral water bottle and used that instead.

The family were obviously distressed but were thankful to finally have their son. In the following days they voiced their anger at what they'd gone through. Darren particularly was highly critical. Plain and simple, it was a botched operation.

As predicted, almost every news outlet in England covered the story and the Thames Valley Police assistant chief constable at the time, Nikki Ross, made the rare move of issuing a public YouTube apology. I had never met her, but I felt sorry for her because she was made to carry the can. The force then referred itself to the Independent Police Complaints Commission (IPCC) which began a formal investigation.

Soon after the debacle, I met Nikki at the annual Women in Policing Awards, where I regularly presented an accolade. I suggested we have a chat over a coffee. We did and she said she was sickened by what had happened and lamented the disbandment of so many police dive units across the region, which left them inadequately resourced whenever divers were required. As a solution I suggested a Memorandum of Understanding between

SGI and Thames Valley Police in which we would be the force's underwater search team, just as we had with other police forces. We signed the contract, and that year were regularly called to the Thames Valley area to help recover drowning victims.

The IPCC issued its findings the following year, which recorded how the family were made to wait 'hours and hours' without contact from the police. It criticized the 'incivility and lack of professionalism' shown to Ellis's family. In legacy to Ellis, I hope no other family are treated in such a manner or ever have to wait unnecessarily for their loved one to be found. Nikki has since retired but we stay in regular contact and we now have an excellent relationship with Thames Valley Police and its brilliant officers.

Chapter 20

When I sat down to write this book, over twenty-five years had passed since the first road protests at Newbury and Honiton that set me on the unusual path I've been treading ever since. And although a quarter of a century is a long time and many things have changed, in other ways much has remained the same.

Eco-warriors continue to hamper big infrastructure projects and fight for the environment. They are more agile now, with organizations such as Extinction Rebellion able to utilize social media to coordinate groups of protesters in different places. Some methods were different. It turned out that gluing yourself to the roads your forebears tried to prevent being built was a simple and effective way to disrupt large numbers of people and get noticed. Some of the causes changed too. The urgency of the climate crisis became the overarching call to action.

But more 'traditional' causes and methods remained. In the UK, at the start of the 2020s the cause célèbre for many environmentalists was HS2, the controversial high-speed railway line stretching from London to Birmingham and the North West. To delay construction, protest camps complete with tunnels dotted the route and some of the old faces turned out to lend their support to the new generation of protesters, one of whom was Swampy, who over the years I'd grown friendly with. We had a lot of shared experience, and we respected each other. We occasionally spoke on the phone, swapping old war stories and sharing opinions on the latest protests. Although we came from different sides of the conflict, the protesters trusted me, we had the same aim – to keep everyone safe.

HS2 marked a change of direction in terms of how protests were policed. As a well-respected adviser to many major infrastructure projects, I was often consulted in the planning phase with regards to how protester management should be run. A person from the Department of Transport and a security consultant for HS2 came to the SGI HQ for advice on how to develop the protest management part of the project. They showed me a report that had been prepared by a large expensive consultancy firm contracted to comment on the likelihood of any protest action on the HS2 route. In summary, the report stated there would be 'little or no protest against HS2 as the rail line cut through swathes of British countryside.' Clearly the person who had written it had never been near a protest camp.

I was cynical. I took the group for a walk around our industrial units, showing them all the specialist equipment that was required to safely carry out these operations. During lunch I gave a presentation and served up a wakeup call. I showed them just how complicated and extensive past protests had been and listed all the ways eco-warriors will put spanners in the works. Jaws dropped. One delegate went white. A lot of memories had dimmed in the twenty years since the UK's road-protest boom of the 1990s. Some people involved in planning HS2 were just children at the time and had no recollection of just how vast some of those protests had been.

'Have no doubt, there are going to be protests, and they will be huge, well-organized and designed to cause maximum disruption,' I said. 'You need to plan for this.'

I continued to offer my advice and guidance to the HS2 project as plans moved forward. I was asked about security and advised that the most cost-effective solution would be to employ one protester removal company for the specialist work and another security company to organize site security. In my opinion there was no one-size-fits-all solution. Safe protester removal was a complex, specialized job requiring specialist tools and know-how. It wasn't something that a security firm would do with a bit of hired muscle and a set of bolt-cutters. The implications of employing under-skilled personnel were obvious to me. There was a real risk of someone with no experience inadvertently collapsing a tunnel with people in it or tearing off an arm moving a concrete barrel that a protester was locked into.

Against my advice, however, the contractor advertised for a single operator, which seemed to me like a massive waste of taxpayer's money because a security firm would still have to subcontract the services of a third party for protester removal and would obviously add their percentage on top. At the tender stage SGI were approached by three companies bidding to provide them with a protester removal solution.

It didn't take long for the protests to start. A group calling themselves HS2 Rebellion secretly dug a hundred-foot tunnel network right outside the headquarters of HS2 in Euston Square Gardens, by Euston Station. They stored up enough food and water supplies to last for weeks and in January and February 2021 occupied the tunnel for thirty-one days, digging and shoring up the structure they had created.

The group included Swampy and another high-profile protester, Dr Larch Maxey.

I watched it all unfold with interest and was asked for my opinion by several news channels. When the bailiffs eventually went in, I was concerned by some of their practices. Clearly, there were no standby rescue teams on site. They were using heavy plant equipment near the entrance of the tunnel. There was no air supply to the protesters and no communications between the parties.

Larch and Swampy rang me from the tunnel to voice their concerns. Then videos started to emerge of bailiffs stamping on protesters' hands.

I rang a contact at the Health and Safety Executive and was clear.

'This needs to stop, someone's going to die on this site,' I told them.

That night the HSE stopped the removal operation and ordered in a standby rescue team.

The protesters' legal team then requested that I visit the site to assess it, but I was warned not to go near the site in a phone call by a very angry HS2 security manager. The last tunneller eventually left of his own volition. All of them had been arrested and later in the year the 'Euston Square Six' faced charges of aggravated trespass at Highbury Corner magistrates' court for their thirty-one days underground. Larch faced a separate charge of damage to a mobile phone. The judge dismissed all the charges and they all walked free from court.

That was just the start. There were several protests along the route of the line as work progressed, hampering progress and hiking up costs. But instead of a considered approach to security and protester removal the police were sent in, effectively using them as a security firm for a private company. In my opinion it was scandalous. Evictions on private land are civil trespass matters and the police should only be involved if there is a threat to life, criminal damage or the enforcement teams have lost control of the situation. Police are being drawn away to deal with civil matters when they should be dealing with serious criminals. While I'm not sticking for one side or the other, the response should be proportionate. Civil matters should be dealt with by private security.

Meanwhile, the original estimate for security on the project rose exponentially and was expected to exceed £200 million.

I had been growing increasingly exasperated by the number of needless drownings and when, in 2019, I was called by Kent Police to search for the body of a six-year-old boy who had fallen into the River Stour in Sandwich, Kent, I was determined to do something to stop the preventable loss of life I witnessed each summer.

Lucas Dobson was the youngest drowning victim we had been called to search for and at the time he was the same age as my daughter, Summer. He was a happy boy with a cheeky smile. His father, Nathan, owned a twenty-five-foot boat and the pair would go out on the river regularly. On 17 August they were walking onto the boat, which was moored, when without warning Lucas tried to jump across to another boat where other children were playing. He did not clear the gap and dropped into the water. The current was so strong that Lucas was swept away the instant he went in. Nathan and three other men jumped in to try to save him, but they struggled themselves and were unable to find him.

At the inquest into the death, Nathan explained tearfully: 'I wanted to see him come up. He didn't come up once. No one saw him, he got sucked out like a bath plug – it just sucked him in.'

We were called in to join the search for the youngster, but the river was fast-moving, tidal, full of reeds and washed back and forth throughout the days. It was impossible to pinpoint a body in the conditions. We recovered Lucas four days later when a police officer spotted his body from the riverbank, not far from where he'd gone in.

In a statement outside the inquest the family paid tribute to their lost child.

'Lucas Dobson was a beautiful, popular, vibrant six-year-old boy. He loved singing, dancing, cars and bikes. His loss has had a devastating effect on his family, so much so that Lucas's mother Kirsty could not even face being here today,' they said.

'The events which led to Lucas's death were entirely preventable. The family are devoting all their efforts to ensuring that Lucas's death brings greater awareness to safety on the river, greater care and attention is taken when children are present and that buoyancy aids are always used.'

After decades of recovering drowning victims and witnessing the devastating impact on the families, I felt it was the right time to put my experiences to positive use and do something to reduce drownings and increase water safety awareness. Nathan's idea was to put lifebuoys along the River Stour and Lucas's mum, Kirsty, wanted to bring in a law to make life jackets compulsory for children.

I devised a plan to create a nationwide water safety scheme through which schools could loan out life jackets to pupils, much like a library book. The idea was that each participating school received a batch of eight life jackets, which parents could request to borrow if they were going near open water. The Lucas Dobson Water Safety Campaign was set up in collaboration with Baltic Lifejackets, who agreed to supply life jackets nearly at cost and print the logo of the Lucas campaign on each one. Fittingly, the logo had a picture of Lucas on it, driving home the point that a real young boy drowned, and a life jacket could have prevented the tragic event.

With help from a crowdfunding page that raised £13,000, the campaign was launched on 17 October 2019. In my own time and at my own expense, I delivered the life jackets to schools in my helicopter all over the UK. In Cheshire my good friend Debbie Davies and her husband Peter helped. She is a star of *The Real Housewives of Cheshire* and a psychic medium. The kids were always excited to see the helicopter land in their playing field and I then delivered a water safety talk. The incredible feedback from the pupils, parents and teachers has been the motivator to continue fundraising and delivering as many life jackets as I can.

Water safety is close to my heart, and I am determined to do all I can to improve access to free life jackets and reduce drownings. Water safety is not part of the national curriculum, which is something that I intend to change. One day, I would like every school in the UK to have free life jackets on offer and a water safety awareness video become mandatory watching.

Sometimes the wheels of justice move easily and the system works as it should, as was the case when carer Samantha Byrne discovered a blood-soaked body in November 2020.

She had arrived at the Surrey home of one of her clients,

Nigel Chapman, and was puzzled by a scribbled note on the floor by the front door addressed to her and her sister – also a carer – which read:

Sorry Kelly/Sam, too early to call you.
Gone to the park to feed the ducks.

It was 7.45 a.m. and Samantha couldn't recall Nigel ever feeding the ducks early in the morning so she was immediately suspicious. She opened the door with her set of keys and noticed an overpowering smell of bleach.

She called out for Mr Chapman, but there was no reply. She went to his bedroom and was horrified to find blood splashes staining the walls and carpet. When she looked down, Mr Chapman was lying motionless on the floor, face down in a pool of his own blood, naked from the waist up. He was dead and had been stabbed repeatedly in the face, chest and neck. A post-mortem found he died from two stab wounds to the heart and discovered superficial stab wounds to the neck and face, as well as defensive wounds to the arms and hands. Several different knives had been used in the attack.

Suspicion fell on one of his alcohol-dependent friends, Jennifer Gwen Lloyd, who had been caught physically beating Nigel several months before. One witness saw her stamping on his chest and stomach and attempting to force a toilet roll down his throat.

As the murder investigation quickly ramped up, a witness came forward and CCTV footage was obtained that showed Lloyd leaving Nigel's home on the day of the murder and dropping the note.

SGI became involved when the police search adviser called our control room to request our assistance a few hours after Nigel had been discovered. We were asked to search a nearby riverbank and the river for any discarded evidence. The water was shallow, so we used a bathyscope, which is an underwater viewer. We quickly found Nigel's keys and his credit card. Lloyd had also dumped a bag that contained his clothes, two knives, empty bottles of bleach and the back door key to his house. She was arrested on suspicion of attempted murder that afternoon. She denied the stabbing and claimed that Nigel had planned to marry her. She also became aggressive and abusive towards the officers, assaulted them and had to be restrained.

A year later, in November 2021, Lloyd was found guilty of murder and sentenced to a minimum jail term of fifteen years and 162 days.

It was a satisfying case to be involved in because we were able to get a positive result quickly that helped ensure that justice was done. The weight of evidence was so overwhelming that halfway through the trial Lloyd changed her not guilty plea to guilty. Channel Five made a documentary about the case, *Killer at the Crime Scene*, in which my team were featured.

But sadly, cases dealt with so swiftly are few and far between. The justice system creaks along. And what of the cases where things go wrong? Where investigations are botched, or where the leads aren't followed? What happens to the families then?

In recent years I have become involved in two of the UK's most infamous cold cases, those of Nicola Payne and Helen McCourt.

In these cases, I helped the elderly parents in their search for the remains of their missing daughters, both of whom were murdered. In each, justice was absent, but neither was it the primary motivation for the parents. Indeed, in Helen's case the murderer had already been convicted, jailed and even released. What the families really wanted was closure and the peace of mind in knowing that their loved ones were properly laid to rest. That luxury had been denied them by the killers.

Without a body, families are in an awful limbo. While both Helen's and Nicola's families accept that their girls are dead, they have nowhere to pay their respects, no grave to tend, no place to sit and talk to their lost loved ones, or just be near them.

I understood this maybe more than others, given my work through the years. I have the luxury of having my father nearby, of knowing there is a place I can go to be near him. These families had nothing, and the parents particularly were haunted by the thought that their daughters were unceremoniously discarded somewhere wrapped in plastic or old blankets, dumped like rubbish in a shallow grave or tipped into a lake.

At the time of writing both cases remained unresolved, but I vowed to carry on searching until all avenues had been exhausted.

Helen went missing in February 1988 aged twenty-two. She disappeared just 700 yards from her home in Billinge, Merseyside, triggering one of the biggest missing person inquiries in the country. Local pub landlord, Ian Simms, was convicted of her murder due to overwhelming forensic evidence. He refused to say where Helen's body was.

Helen's mum, Marie, wrote to Simms in jail in 1992 begging him to reveal where he'd put her daughter. He refused and his reply was full of threats.

Marie campaigned for Helen's Law, which makes it harder for

killers to get parole if they refuse to reveal where they hid their victims. Simms was released in February 2020 after serving thirty-two years in jail, before the law was enacted, and never gave Marie the peace of mind she deserved. All she wants is her daughter back.

I became involved in the search after the family approached me and asked for my help. For several years I analysed the evidence and went back over the case to identify areas that were missed in the original investigation, or that needed to be reinvestigated with the benefit of new technology. At the time of writing, I am still involved in the case, and continue to search a specific area. I am confident that Helen is buried within 200 metres of where I have been searching.

Nicola was eighteen when she went missing on Saturday, 14 December 1991. Her son was just six months old and she was looking forward to spending her first Christmas with him. She had been staying with her boyfriend near the Potters Green area of Coventry the night before she disappeared, and that morning set out to walk across a small area of wasteland known locally as the Black Pad that provided a convenient shortcut to her parents' house a few hundred metres away. She never arrived.

The search for Nicola involved more than eighty police officers, sniffer dogs and helicopters with heat-seeking cameras. In the days that followed her disappearance, several witnesses described seeing two men acting suspiciously in the area at the time. They also described seeing a distinctive blue Ford Capri nearby. There were several arrests but there was no body and the case proved hard to crack. Generally, the longer a crime goes unexplained, the harder it is to solve. Leads go cold, evidence degrades, disposition sites grow over, remains decompose, memories fade.

In Nicola's case new information led to new searches. In 1996, a garden was excavated. In 2001 part of the Oxford Canal near Ansty was dredged. In 2005 Nicola's family made a fresh appeal for information. In March 2007 the case was reviewed. When Helen's son was sixteen, he issued a statement appealing for information. In November that year a man was arrested and questioned but released without charge. In June 2008 another garden was excavated. In June 2012 more land was dug up and searched and two men were arrested but soon discounted from the inquiry. In December 2012, on the twenty-first anniversary of Helen's disappearance, Nicola's parents made another appeal. Two years later, in March 2014, police began searching Coombe Park fishery. In May 2014 two of the original suspects were rearrested and released on bail. In January 2015 they were charged with Nicola's murder and in October 2015 they were tried but found not guilty.

The defence described the police investigation in the early 1990s as 'sloppy', highlighting discrepancies in the way evidence was stored.

In 2016, after a witness came forward to say they had seen two men acting suspiciously by the lake in Coombe Country Park on the day Nicola disappeared, police conducted a twelve-week search of the area. It was unsuccessful.

Three decades after she went missing Nicola remained lost to her family and her killers were still free.

It was believed she had been abducted and killed and her body dumped in a lake or canal near to where she went missing. Her parents, both exhausted and in poor health, had headed the campaign to find her and handed the baton over to Nicola's brothers and cousin. There was also a £100,000 reward for information leading to the recovery of her remains.

I became involved in the case through a contact, TV presenter and former detective Mark Williams-Thomas. Together we did a deep dive through the evidence, worked out search strategies and once again started looking over areas that featured in the original and subsequent investigations. I used both GPR and SSS and, as in the case of Helen McCourt, I believe that we are narrowing down the search area and will eventually find Nicola, or at least discover what happened to her.

For families facing this agonizing limbo of uncertainty, discovery of a body will inevitably be bittersweet, bringing closure but confirming their worst fears. As I was writing this book one historical search I had been involved in reached a sad conclusion. In February 2019 I was contacted by the Thames Valley Police PolSA and asked to help in the search for missing student Leah Croucher. I was called just days after the nineteen-year-old was last sighted and asked to search various lakes and disused quarries, particularly the Blue Lagoon Nature Reserve in Bletchley. The searches went on for several months.

Leah was last seen on CCTV walking on Buzzacott Lane in the Furzton area of Milton Keynes at 8.16 a.m. on 15 February 2019. I knew if Leah was in the water, I would be able to locate her quickly using our side scan sonar and recover her using my underwater search team. When the side scan sonar was deployed, I sat in the boat intensely monitoring the crystal-clear images on the screen. I could see every stick, every boulder, every detail on the lakebed, but no Leah. I always want a quick result for the family, either to rule out her being there or to confirm their worst fears. Ultimately my aim is to bring some form of closure, no matter how terrible that may be.

I met Leah's family at the nature reserve and knew in the back of my mind that their daughter would not be found alive. I realized the terrible pain they must have been going through.

Between February and October that year we searched various lakes around Milton Keynes but could find no trace of Leah. The family would have to wait another three years before they discovered the awful truth about what happened to her. In October 2022, human remains were discovered in a loft space at a detached house in Furzton and were officially identified as hers. Thames Valley Police said an initial post-mortem examination had been unable to establish the cause of death and further investigations were continuing.

Soon after the search for Leah, we were called to help search for another missing teenager. In 2020 I was asked by West Midlands Police to help search for Natalie Putt, who disappeared in September 2003. My expertise was sought and I carried out a number of covert searches at various locations. But, as I write this, there is still no sign of Natalie and the family remains unable to move on.

When these historical cases were first investigated there were no specialist police search teams and forensic awareness was not as well-drilled as it is today. The one thing that can bring these families hope is that technology is developing all the time. Even commonly available open-source technology is now being used to solve cold cases and make searches easier. For example, several years ago I was asked to search areas of interest that had been identified as possible disposition sites in the murder case of Linda Razzell, who went missing on her way to work at Swindon College in 2001 and who had never been found.

Her husband, Glyn, was convicted of her murder. At the time the couple were involved in divorce proceedings. He continued to plead his innocence behind bars and enlisted miscarriage of justice campaigners. His case featured in a 2018 documentary *Conviction: Murder in Suburbia*, but his campaign backfired when the programme failed to turn up sufficient leads and parole officials used it as grounds to deny release because it upset Linda's grieving relatives. In a later attempt to get parole he was denied freedom again under the terms of Helen's Law because he refused to say where he'd put Linda's body.

I was asked to help when the case was reviewed, and searched air shafts, a wood and areas of open water. In one air shaft I found a bag of bones, but they were badger remains, no doubt dumped there by someone who had been illegally badger baiting. As the case was so old and the areas where we were tasked with searching had changed considerably, I was able to use historic satellite imagery from Google Earth to wind back the clock and get an accurate view of what the area was like at the time of Linda's disappearance. This allowed me to build a grid to search and identify possible disposition sites based on the topography of the area as it appeared in 2001.

Google Earth is one of the many easily accessible tools available to investigators today that can help with cases. For example, in the search for Linda Razzell, whose body has never been found, I was able to map out the footprint of a public toilet that was a feature of the area at the time but had long since been pulled down.

There are many other missing body cases that I've been

involved in where eventually technology may unlock the puzzle, and the growing interest in crime and forensics hopefully will mean that one day the McCourt family, the Payne family and all the other families suffering the awful heartache of a missing loved one will get closure.

Training is important, too, and I regularly host groups of forensics students on my farm where I give talks and demonstrations of the latest kit. When some of our rescue animals meet their natural ends, I bury them on parts of the farm so I can study what happens to their graves and let students practise using the GPR over their resting places. There are several alpaca bodies dotted around.

I've been incredibly fortunate to have led the life I have. I was written off at school and only had a vague idea of what I wanted to do, but I persevered and I worked hard and when I found what I was passionate about, I stuck at it and stayed positive when there were setbacks.

From very humble beginnings, I founded SGI and the company has gone from strength to strength and excelled in some very challenging situations. It is now a world-renowned organization with a highly professional and disciplined team, capable of carrying out dangerous and complex operations on land and underwater. To all intents and purposes, we are the real-life Thunderbirds.

We now have contracts with a number of police forces for underwater search and recovery operations.

I've worked with some of the best law enforcement, emergency services personnel and other specialists in the field who dedicate their lives to the greater good. I have the utmost respect for all of them and the work they do.

I've been lucky enough to be able to share my expertise and knowledge not only with students but also with the wider public through television work and interviews. I advised author Peter James about search techniques for his crime novels and producers of the television drama *Call Red* about underground rescue. (I even had a cameo role in the show.) I led my team on the Sky reality show *The Heist*, in which contestants 'stole' money and had to hide it. Those that managed to keep the whereabouts of the swag secret got to keep it. Our job was to locate it, which we did, much to the contestants' dismay. I continue to appear on many crime podcasts and documentaries.

I've been supported throughout by the best and most loving family anyone could ever wish for and I'll always be grateful to them. I'm happiest on my farm with my loved ones and my animals. Sometimes my professional life can be grim – there's no getting away from that. But my home is my sanctuary. It is where I can go to be myself and to relax and decompress.

All these professional add-ons, however, are nothing in comparison to the real work I do. From rescuing people from the mines, removing protesters from tunnels, searching for buried human remains or methodically searching the bottom of a lake for a missing person, the knowledge that I can bring closure to a grieving family tells me that I am right where I should be, searching in the shadows, under the surface. I am proud of what I have achieved but none of it would have happened without my dad taking me down the mines as a child. I miss him so much and wish he was here to see what I've accomplished. From the very first moment I went into that dark tunnel on my own, aged five, I knew where I was meant to be.

Acknowledgements

I have to start by thanking my wonderful parents John and Nora for their endless support, instilling in me a sense of adventure and encouraging me to follow my dreams and never give up. My life has seldom been lived in the comfort zone and they would have had many sleepless nights.

I am enormously grateful to Nick Harding for his patience, guidance and humour while helping me write my story – many hours and many laughs were had.

Thank you to editor Lydia Ramah, as well as Helena Caldon, Sian Chilvers, Sara Cywinski, Samantha Fletcher, Ross Jamieson, Holly Sheldrake, Natasha Tulett, Josie Turner, Stuart Wilson and the rest of the brilliant team at Pan Macmillan for publishing my book and allowing me to tell my story.

During my life I have met so many dedicated and professional people – far too many to mention. From police, emergency services, military in the UK and around the world who dedicate their lives to the greater good, I thank you. And to all my great friends I made in the Parachute Regiment.

To my brilliant team at Specialist Group International, who carry out some harrowing work in all weathers around the clock, to save lives and help bring closure to families who have lost loved ones.

To my family and friends in the UK, Canada and Australia.

To my beautiful daughters, Summer, Natasha and Danielle. I'm immensely proud of you all. Thank you for being there for me. And finally, to my soulmate Adele, for your love and encouragement. Thank you for being there every step of the way. We make a great team.